The Art of Uncertainty

David Spiegelhalter

THE ART OF UNCERTAINTY

How to Navigate Chance,
Ignorance, Risk and Luck

W. W. NORTON & COMPANY
Independent Publishers Since 1923

First published in Great Britain by Pelican Books in 2024.

For information about permission to reproduce selections from this book, write to Permissions, W. W. Norton & Company, Inc., 500 Fifth Avenue, New York, NY 10110

For information about special discounts for bulk purchases, please contact W. W. Norton Special Sales at specialsales@wwnorton.com or 800-233-4830

Manufacturing by Lake Book Manufacturing

ISBN: 978-1-324-10611-1

W. W. Norton & Company, Inc.
500 Fifth Avenue, New York, NY 10110
www.wwnorton.com

W. W. Norton & Company Ltd.
15 Carlisle Street, London W1D 3BS

1 2 3 4 5 6 7 8 9 0

To chance, luck, fate, destiny, karma,
the goddess Fortuna, or whatever it is that gave me
my wonderful partner, family, friends and colleagues

Table of Contents

List of Figures

List of Tables

Introduction

'I returned and saw under the sun, that the race is not to the swift, nor the battle to the strong, nor bread to the wise, nor riches to men of understanding, nor favour to men of skill; but time and chance happen to them all.'

—Ecclesiastes 9:11, *New King James Bible*

It was 29 January 1918, and the 35-year-old gas officer for British 104th Brigade, Ypres section, started off on another day's inspection duties north of Passchendaele on the Western Front. The appalling battles of the previous year in that area, which cost around 250,000 casualties on each side, had left a desolate landscape of mud and ruins. His routes were by roads and trenches that had been pinpointed by German artillery, and in the six weeks he had been in post his diary had variously recorded: 'Narrow escape on return journey', 'Lucky to get through in time', 'Artillery strafe'. On that day he visited Eagle trench near the front line, but on the way back he was, as he termed it later, 'blown up', and ended up in a field dressing station. He was then taken by ambulance to No. 64 Casualty Clearing Station. It appeared as if his luck had run out. But had it?

This gas officer was my grandfather, Cecil Spiegelhalter, and ironically his experiences that day turned out to be very fortunate. After his injury, he was categorized as unfit for front-line duties and spent the rest of the war well behind the lines. Meanwhile, his old battalion, 18th, Lancashire Fusiliers, was moved to the Somme, a supposedly quiet area after the battles of 1916 which cost around a million casualties. But they were just in time to face the massive spring offensive of 1918, fighting a desperate rearguard action and then going over the top in vain attempts to recapture territory.

He had been lucky the previous year as well. He had been made a second lieutenant, the most dangerous rank in the army, requiring him to be first up the ladder in an attack and to encourage his men to follow. But he got a severe fever and was recuperating at Thirtle Bridge Camp in Yorkshire while his battalion was engaged in some of the worst fighting of the war.*

Of course, if one of those shells had fallen a bit closer, or if he had had to lead his men into attack, I would probably not be here to tell the story. And this is only one of the long chain of fortuitous events that led to my existence: my mother being captured by pirates in the South China Sea and later evacuating from Shanghai in 1937 under shellfire; my parents meeting in the war; my father closely avoiding plane crashes in the RAF and then nearly dying of tuberculosis. Then when a cold snap hit the UK in November 1952, they were living in a barely heated stone cottage, with no television and nothing to do but go to bed early to keep warm . . . and here I am.

I don't think it makes much sense to try to assess a numerical probability of each of us being born. All we can say is that, like everything else that happens in this immeasurably complex world, each of us is the result of an unforeseen and unforeseeable sequence of small occurrences, or 'micro-contingencies'. But what underlies and drives this fragile chain of events?

Our feelings about this question depend on our philosophy, and even our spiritual beliefs. Terms such as *fate, destiny, fortune, God's will, karma* suggest an underlying cause or even predestination, while words like *chance, happenstance, vicissitude, luck* bring to mind capricious and uncontrolled randomness. This is deeply personal, and not generally an area of rational thought,

---

* My grandfather was on the British side. Although his father had been an indentured economic migrant from Germany, Cecil was a proud Yorkshireman who saw no need to anglicize his name to Salter, as other Spiegelhalters had done, and his ancestry came in useful when he taught German in the British Staff College in Cologne in 1919. After being blown up, he was classified as having 'Disordered Action of the Heart' (DAH), also known as 'soldier's heart', a condition with acute symptoms but no apparent organic cause. It later became known as shellshock. Incidentally, a fellow officer at Thirtle Bridge Camp was a certain J. R. R. Tolkien, who was also recovering from his trench experiences in 1916.

as in the old gamblers' saying: 'It's unlucky to be superstitious.' But whatever our worldview, Cecil's experiences reflect the essential uncertainty in our life, both in our coming into existence in the first place and, subsequently, what happens to us, and the world.

This constant state of uncertainty is an essential part of the human condition. It may be mundane ('what's for lunch?') or existential ('will there be a catastrophic global war in the next hundred years?'), and the degree of uncertainty in our lives can vary considerably across time and place. All through history the vast mass of people have lived repetitive and unchanging lives, starved of opportunities to develop and change, although they were certainly not free from risk. Cultural historian Jerry Toner has pointed out that life expectancy in the Roman era was around twenty-five years, with hunger, cold, disease and violence all taking their toll.[1] While on a global scale, some periods have been particularly volatile and may justifiably deserve labelling as 'ages of uncertainty'; for example, the 1930s saw an extended period of rising national and international tensions, eventually leading to massive conflict.

The recent pandemic revealed many vulnerabilities in society, and populist pressures on democracy, wars and conflicts, climate change and other global threats may make us feel we are currently living in another age of uncertainty. And there are domestic challenges as well: an Ipsos 'What Worries the World Survey for December 2022'[2] asked 'Which three topics do you find the most worrying in your country?', and the top six responses were inflation, poverty, crime, unemployment, corruption and the quality of health care. Inflation has recently come into prominence, and Covid-19 was a temporary inclusion, but these six have been the usual suspects over recent years.

In some respects, our lives may also seem less predictable than those of previous generations. Although there were more childhood illnesses, and life expectancy was considerably shorter than today, when I was young I could access student grants, state education, health care and other post-war innovations in the UK, and I had an unconscious assumption that I could get a reasonable job then stay in it for years if it suited me – which is precisely what happened, as I worked for the Medical Research Council for thirty-two years. Now, job insecurity and the gig economy are the norm. Twenty-one per cent of millennials (born 1981–96) say they've changed jobs within the past year, more than three times the rate for non-millennials.[3] Such uncertainty may not, however, necessarily be only a negative condition – starting a new job in an unknown organization might be experienced as both a source of anxiety and a major opportunity.

Uncertainty is all about us, but, like the air we breathe, it tends to remain unexamined. This book will try to do something about that.

———————

I've spent my whole career working on investigations aimed at reducing uncertainty about what is happening, what might happen, and even the reasons why things happen. This has generally involved examining masses of data and assessing what we can learn from the available evidence. This book has arisen from my own experiences in trying to judge, and then explain to others, how much confidence we might have in claims made in the face of uncertainty.

All this work has taught me one major lesson, which runs through all the ideas, questions and stories in this book. Put simply, it's that uncertainty is a *relationship* between someone (perhaps 'you') and the outside world, so it depends on the subjective perspective and knowledge of the observer. Our personal judgements therefore play an essential role whenever we are faced with uncertainty, whether we are thinking about our lives, weighing up what people tell us, or doing scientific research. Again, tolerance of uncertainty can vary hugely between people – some might get a sense of excitement from unpredictability, while others feel chronic anxiety.

But just because uncertainty is personal, it does not mean it is just about feelings. In *Thinking, Fast and Slow*,[4] psychologist Daniel Kahneman popularized the idea of two systems of thinking – one is rapid and intuitive, and the other more considered and analytic. When it comes to dealing with uncertainty, he argues that the first, fast system tends towards over-confidence, neglects important background information, ignores the quality and the quantity of evidence, is unduly influenced by how the issue is posed, takes too much notice of rare but dramatic events, and suppresses doubt. These are not characteristics that should be encouraged.

In contrast, this book is about trying to think slowly about our 'not-knowing'. Such an analytic approach should not only bring some clarity to our own situation, but also empower us to judge if anyone – whether a politician, journalist, scientist, or some influencer on social media expressing complete certainty in their bizarre beliefs – is being far more confident than they should be.

———————

As befits a book about uncertainty, I shall focus on what could, in theory at least, become certain. This may sound obvious, but means I can neatly avoid dealing with personal doubts about, for example, the best Beatles song, or what to wear this evening, or the existence of God. These are not verifiable 'facts' and so, while we might say we are 'uncertain', we are really expressing opinion, indecision, or faith, which (fortunately) are outside my remit.

With that understanding, here is a quick overview of how we shall deconstruct the idea of uncertainty. To start with, our everyday language is full of words like *unlikely*, *possible*, *likely*, *probable*, *rare*, and so on, but these vague terms are easily misunderstood, and we shall see they may have even increased the risk of nuclear war. If we want a firmer grip on uncertainty, we need to start using numbers, and a first step is to try to define what we mean by words such as *likely*. A simple quiz can then show that not only can we put our ignorance into numbers, but also score how good our judgements are – and see how super-forecasters think.

But if it is so useful to express our uncertainty in numbers, why was the idea of probability so late in arriving, even though people had been playing with knucklebones and dice for millennia? It was not until the Renaissance that any attempt was made to analyse what was going on when dice were thrown, and then, like the release from a pent-up dam, the field exploded into applications in pensions and annuities, astronomy and the law, as well as, of course, gambling. Admittedly, the nuts and bolts of probability can be tricky – even school exam questions can be a bit baffling. But it can help us answer questions such as whether, in the whole of history, any two packs of cards have been in exactly the same order after a good shuffle, and understand how Casanova's mathematical skills led to an extraordinarily successful French lottery. Although we have to admit that probability is a very odd thing – there's no measuring instrument for it, so is it an 'objective' aspect of the world, or is it all in the eye of the beholder? Does it really exist at all?

I am often asked 'what's the chance of that?' after something apparently surprising has occurred, and this has led to a personal fascination with coincidences and luck. Probability can help explain why surprising events happen so often, although you may still be amazed by the mystery of Ron Biederman's trousers in Chapter 4, and what about Mr and Mrs Huntrodds of Whitby, whose births, marriage and deaths all occurred on 19 September? What are the chances of a union like that? And was illusionist Derren Brown lucky or unlucky when he flipped ten heads in a row on television?

When we take apart the concept of luck, the most important type turns out to be 'constitutive luck' – essentially, who you were born as. Of course, we can only think about this because we *were* born, and we've already noted the fragile chain of events that led each of us to be in this world. But does the world, including our birth, operate according to complex mechanical laws, or are our lives governed by genuine randomness? I shall try to sidestep this centuries-old question, although, whatever your opinion, there is no doubt that 'effective' randomness is extraordinarily useful, whether to ensure fair allocation, to balance groups given different medical treatments, in taking football penalties, or building an atomic bomb. But are random-number generators, or the UK lottery, truly random?

Once we accept a personal, subjective view of probability and uncertainty, we are led naturally to *Bayesian analysis*, in which we use the theory of probability to revise our beliefs in the light of new evidence. These ideas were crucial to Alan Turing's codebreaking in the Second World War and now help us to interpret imperfect data, such as automated recognition of faces in crowds. We might even have Bayesian brains.

Of course, no amount of new evidence can shift our opinions if we have a closed mind that refuses to admit uncertainty, although, bizarrely, Oliver Cromwell has a lot to teach us about having such humility. Fortunately, there was some humility shown during the Covid pandemic when up to twelve different methods were used simultaneously to estimate the constantly changing infection rate in the UK – a good illustration of the importance of exploring a diversity of views when basing claims on statistical models.

This example also shows that, although scientific investigations are generally fairly good at acknowledging uncertainty, any calculated margins of error will generally be too small, since they are conditional on all the assumptions in the statistical model being true – and it has become a cliché that 'all models are wrong'. We also inevitably have a feeling that some analyses are better than others, as the evidence is stronger and the understanding is better. Pioneered by the intelligence community, many organizations have found it useful to express a level of *confidence* in any analysis, as our team did when estimating how many people had been infected with hepatitis C through infected blood transfusions in the UK.

It's all very well trying to work out *what* has occurred, but there is often uncertainty about *why* something happened, and who or what was to blame. Is human activity behind both the global rise in temperature, and specifically the unprecedentedly warm UK autumn of 2023? Why did a British ship twice the size of the *Titanic* sink without a trace in 1980? In civil courts of law, judges may use probability theory to decide whether exposure to chemicals at work is to blame for an ex-employee's cancer, while criminal trials look for evidence 'beyond reasonable doubt' to make a conviction. Sadly, the stories of mothers who have been wrongly convicted of murdering their children show that probabilities can be misused in courts, when events are claimed to be too unlikely to be just coincidence.

Perhaps the archetypal expressions of uncertainty occur when predicting the future, whether it's the following day's football results, next week's weather, or next year's economic growth, and you may even be interested in how long you might live, and whether global warming will reach catastrophic levels this century. All these predictions require a mixture of mathematical modelling and a large dose of judgement. There's particular interest in the risks of crises and disasters, and we will look at the probabilities given in 1975 to an accident in a nuclear plant killing over a thousand people, and the 2023 judgements of whether the UK government will face strategic hostage-taking and other threats over the next five years.

————

There is one unavoidable quote when it comes to discussing uncertainty:

> as we know, there are known knowns; there are things we know we know. We also know there are known unknowns; that is to say we know there are some things we do not know. But there are also unknown unknowns – the ones we don't know we don't know.
>
> —United States Secretary of Defense Donald Rumsfeld in 2002

This was widely ridiculed at the time, but has since been accepted as an important contribution to the language of not-knowing. Science has generally been concerned with the 'known unknowns' – where we can list the possibilities, construct mathematical models and express our uncertainty in numbers. In contrast, Rumsfeld's 'unknown unknowns' can include *delusions* – things we mistakenly think we know, such as unquestioned (but inappropriate) assumptions in our analysis, or a confident (but inadequate) list of possible future events. One aim of this book is to encourage sufficient humility to turn the unknown unknowns into known, or at least acknowledged, unknowns, and so hopefully avoid being taken by complete surprise. This may require facing up to *deep uncertainty* – limitations of our whole conceptualization of the world, reflecting the boundaries of our ideas as to what could happen. This requires admitting both the gaps in our understanding and the limits to our imagination, and rather than doing yet more complex analysis and trying to produce an optimal course of action, it may be better to seek flexible strategies that should be resilient to most eventualities.

Rumsfeld did leave out one combination – the *unknown knowns*, which philosopher Slavoj Žižek described as 'things we don't know that we know, all the unconscious beliefs and prejudices that determine how we perceive reality and intervene in it'.[5] More generously, this category may include accurate understanding that we do not know we have – so-called *tacit knowledge*.

While huge effort has gone into technical methods for assessing the magnitude of risks, there has been limited attention to the challenges in *communicating* uncertainty. Politicians may exaggerate their confidence, as in the build-up to the 2003 Iraq War, but if we are to communicate in a trustworthy way then we need to be clear about the potential benefits and harms of any decisions, even if it's just to point out how much TV someone would have to watch before they can expect to get a pulmonary embolism.* A frequent excuse for keeping quiet about uncertainty is that it may mean that audiences will lose their trust in the communicator, but we shall see evidence that suggests the opposite may be the case.

We all make decisions in the face of uncertainty, and although in theory there is a formal mechanism to decide the best action, as individuals we tend to mainly use our instincts, perhaps imagining stories about what might happen if we are lucky or unlucky. We expect more of government health-and-safety regulators, who have the delicate task of deciding about 'tolerable' risks to employees and the public, which means that in the UK there is an official 'acceptable' risk of being killed at work. Although there have been problems in specifying how much burnt toast it is safe to consume each day.

Finally, we look into a future of AI, climate change, international instability, and a litany of threats and opportunities. We have to confront the fact that we don't know what we don't know, that our understanding is always inadequate, and that we should genuinely acknowledge our uncertainty. But this basic humility need not stop us from considering plausible futures, making decisions and getting on with our lives.

———

After that brief look at the book's contents, a small apology to those anxious about mathematics. I'm afraid it's impossible to completely avoid technical

---

* Spoiler: five hours a night for 19,000 years. Roughly.

material when discussing probability, but this is kept to a minimum, and can be skimmed over if you prefer. Most workings are in footnotes, to avoid distraction for those who don't want the flow interrupted. A glossary provides definitions and further technical explanation for terms in **bold**, and full endnotes are provided for each chapter.

Terminology can be tricky. Words like **probability**, **chance** and **likelihood** are often used interchangeably in everyday language, but I shall be a bit more pedantic. *Probability* will be kept for numbers expressing uncertainty, although when the probabilities can be generally agreed upon due to shared understanding of the underlying process, such as when flipping coins, I will refer to them as *chances* – I will also use *chance* as a more general term for unpredictability. *Likelihood* will generally be restricted to its technical meaning, explained in Chapter 7. **Risk** can mean almost anything you want it to, and in everyday language is often used to describe both a threat ('that broken paving stone is a definite risk'), and the chance of an event ('the risk of falling over is small'). I will use the term loosely, and let it take its meaning from the context.

This book is intended for a range of readers: students of probability who want to go beyond the standard mathematical syllabus; all those who work in something to do with 'risk' who want to explore outside their specific area; scientists who want to examine further how to communicate both quantified and unquantified uncertainty arising from their work; and perhaps most importantly, the interested citizen who is largely reliant on 'experts' and wishes to assess their trustworthiness.

Unavoidable uncertainty is part of the human condition, and only a minority of people want to know what they are going to get for Christmas, or (assuming it were possible to know), when they are going to die. An explicit and sometimes uncomfortable consciousness of uncertainty is part of what makes us human. Although we may prefer to ignore it, I hope this book may do something to help readers accept, and possibly even enjoy, the experience of not knowing.

Summary

- Our very existence depends on a fragile chain of unforeseeable events.
- We all have to live with uncertainty, about what is going to happen, what may have happened in the past, and how the world works.
- Uncertainty is a relationship, with a subject who considers an object that they are uncertain about.
- We have varied individual feelings about coincidence and luck, and doubts about the future.
- Probability is the formal language of uncertainty, but any application involves a model of the real world dependent on numerous assumptions.
- Probability models are always inadequate, and we may need to acknowledge deeper uncertainty.
- We may prefer to ignore uncertainty, but it would be better to acknowledge it.

CHAPTER 1

Uncertainty is Personal

'There is no such thing as absolute certainty, but there is assurance sufficient for the purposes of human life.'

—John Stuart Mill, *On Liberty*

Flipping a coin is an archetypal example of dealing with uncertainty. Imagine I am standing in front of you with a typical coin in my hand, about to flip it.* I then ask you the probability that it will come up heads. You happily say 'half', or '50%', or '50–50', or 'one in two'.

I then flip the coin, catch it, but cover it up before you see it, although I take a quick peek. I then ask, what's your probability it's heads?

Things have changed, as the event is now decided – there is no randomness left, just ignorance. Not only that, but I know the answer and you don't, which is a situation that some can find unnerving. Most people are now hesitant to give an answer but may eventually repeat 'half' or similar, although somewhat grudgingly.

This simple exercise provides a number of lessons. First, note that I used the term '*your* probability' rather than '*the* probability', emphasizing your role as the owner of the uncertainty, making you the *subject*. *My* probability would be different, being either 1 or 0 depending on whether the coin had landed with heads or tails uppermost. Second, the *object* of uncertainty was

---

* Apologies for reusing this example from *The Art of Statistics*, but I could not think of a better one.

originally the result of a future flip, where the uncertainty is due to what we might call chance, or unavoidable unpredictability; this is sometimes termed **aleatory uncertainty**, about the future we *cannot know*. But now the object is the current state of the coin, and the uncertainty is due to your lack of knowledge; this is called **epistemic uncertainty**, about what we currently *do not know*.

Ancient oracles are popularly thought to be solely concerned with the unknown future, but classicist Esther Eidinow has pointed out that they were more used to try to resolve epistemic uncertainty, being also questioned about the unknown present and unknown past; for example, the oracle of Dodona was asked about 'who stole sheepskins, who stole the silver or who murdered somebody'.[1]

Throughout the book we shall continue to explore uncertainty both about what we cannot know (yet), and what we don't know (but possibly could). But now we are ready to attempt to answer the crucial question:

What is uncertainty?

Most formal definitions say it is a 'lack of certainty', so we need to look at definitions of 'certainty'. The consensus is along the lines of

Certainty: firm conviction, with no doubts, that something is the case.

This clearly expresses the idea that certainty is a personal feeling. Therefore so is uncertainty, which occurs when someone does *not* have firm convictions and *does* harbour doubts. This is reflected in a more formal definition,[2] which I personally find appealing:

Uncertainty: the conscious awareness of ignorance.

The crucial issue reflected in these definitions is that (with possible sub-atomic exceptions, as we shall come to in Chapter 3) we shall not be thinking of uncertainty as a property of the world but of our *relationship* with the world. This means that two individuals or groups can, quite reasonably, have different degrees of uncertainty about exactly the same thing, due to them having different knowledge or perspectives, as we found with the spun coin.* This vital idea will run through the whole book.

Once we accept that uncertainty is a relationship, we can explore its possible characteristics. These include:

- A *subject* who is uncertain, whether an individual person or a consensus of a group, so ideally we should always say *my* probability or *your* probability, or whatever is appropriate. Although I shall tend to use *the* probability, or *the* chance, when there would be general agreement, say because of a clear underlying physical mechanism, such as flips of a coin, lottery draws, or dates of birth.

- An *object* that they are uncertain about, which can be any aspect of the world that is, at least potentially, verifiable; say about what has gone on in the past, what is happening at the moment, accepted facts, how things work, what caused what, and what the future may bring. As mentioned in the Introduction, this focus on well-defined objects means that we will avoid dealing with the many looser ways in which the term 'uncertainty' is used, such as free-floating ideas of unease, unverifiable claims about whether there is a God or not, *indecision* about what to do, or *imprecision* arising from vague language.

- A *context*, in terms of what is known or assumed by whoever is uncertain – this becomes vital when we start thinking about statistical models.

- A *source*, in terms of what drives the uncertainty – this can include natural variation in anything we want to measure, inherent 'randomness' of nature, differences between people, limited knowledge, ambiguous information, sheer complexity that limits comprehension, limitations in computation, the possibility of errors, or just irreducible ignorance of what is going on.

- An *expression* of the uncertainty, whether verbal, numerical, or visual, which will generally communicate some idea of magnitude, and will be based on background understanding and assumptions.

---

* I will skate over whether one can ever be absolutely certain of anything at all. I have a running joke with colleagues that whenever we come to the question 'What is truth?', it is time to give up and change the subject. Similarly, I am not going to try to say what is 'truly certain', and just acknowledge that we can strive to get greater understanding of how things actually are and so become 'certain enough' but never reach complete understanding. See the quote from John Stuart Mill at the start of the chapter.

- Where appropriate, an *emotional* response to the uncertainty, also known as the 'affect', which might be dread, excitement, anxiety, resignation, and so on, and can have physical manifestations of 'butterflies in the stomach', upset sleep, and so on.

In the example of the flipped coin, the subject is you, the object is the result of the coin flip, the context (after sensibly checking that I have not substituted a double-headed coin) is that the coin is assumed fair, the source of uncertainty is the fact that I have spun the coin and covered up the result, the expression is a numerical probability, and the emotional response may well be irritation.

The spun coin is an extreme example of asymmetry of information – when one participant knows more than another. But, as the following story shows, even small asymmetries in uncertainty can prove very rewarding (if you can get away with it).

How do you win £7.7 million at cards?

The simple answer is – you cheat. At least this is what the UK Supreme Court* decided, after star poker-player Phil Ivey won this amount at Crockfords casino in London in 2012. As recounted in the court judgement,[3] Ivey was playing Punto Banco, a form of baccarat in which cards are drawn from a 'shoe' of eight shuffled packs and each player tries to get as close as possible to a total of nine. After a new shoe was started at 9 p.m. on 20 August, Ivey and his companion started making unusual requests that certain cards were placed back in the shoe in a specific orientation, claiming this was 'for luck'. By 10 p.m. the shoe was exhausted, and he asked that the same shoe be used again as he 'had won £40,000 with that deck', and also requested that the cards be not shuffled by hand but in a machine (which was guaranteed to preserve their orientation). Ivey had won £2 million by 4 a.m. on the 21st and asked that the same shoe be retained for when he returned to play again. He came back at 3 p.m., and by 6.40 p.m. had won a total of over £7.7 million.

---

* I wrote about this case in the *UK Supreme Court Yearbook*, admittedly an unusual place for a statistician to publish.

Understandably, Crockfords were suspicious, but could not immediately see what he had done.

Card-counting, in which a player keeps a tally of what cards have been played, is legal (although frowned upon by casinos), but it is not much use in Punto Banco. Rather than card-counting, Ivey was *edge-sorting*. Cards have a particular pattern, often a grid of circles, on the back, and in production of a pack the left and right edges may not intersect the pattern identically, meaning that the orientation of the card can be judged from the back. Ivey tried to ensure that important cards (specifically important in this game are sevens, eights and nines) were replaced in one orientation, so that they could be identified from their backs when they reappeared in the shoe. The Supreme Court reckoned he changed the balance of the game from an advantage of over 1% to the casino to over 6% in his favour.

Crockfords, rather remarkably, realized this only after scrutiny of video footage. They refused to pay up, and Ivey took them to court, where he freely admitted his strategy but said he did not consider it cheating but just 'advantage' play, similar to card-counting. His lawyers argued that he was not being dishonest according to accepted English law, which requires that the person know their actions are dishonest. The case went all the way to the Supreme Court, which ruled against Ivey, fundamentally changing the legal criterion for dishonesty to requiring only that a 'reasonable person' would think the actions dishonest, regardless of the perpetrator's own perception.

This example demonstrates the subjectivity of uncertainty. Ivey's actions did not in any way change the randomness in the order of the cards – the 'chances' of each card appearing were unaffected. But, after scrutinizing the pattern on the back of the cards, he did change his degree of personal uncertainty about what the next card would be, and so he could adapt his betting – the cards were still, strictly speaking, 'unknowable', but they were slightly less unknowable to him, as his interference had produced a notable asymmetry in knowledge between him and the casino. But not only did he lose his winnings, he faced massive legal costs.

––––––––

Try this thought experiment:

- *Note the time and date, and close your eyes. Think of what you will be doing in 1 minute. Now try 1 hour, 1 day, 1 week, 1 year, 20 years.*

For short periods we have a good idea of what might happen, but as the horizon lengthens, the future possibilities spread outward like strands of spaghetti. We cannot even imagine all the possibilities, let alone know which one we will take, and whether, like my grandfather, we will avoid the (hopefully metaphorical) exploding shells.

- *Now try to remember – what were you doing exactly 1 day ago, 1 year ago, 10 years ago?*

This is rather different from trying to look into the future. We could in principle find out what has happened in the past, but we can't immediately recall the specific chain of events that brought us to where we are now – most of the past soon fades into a rather indistinct blur. As we saw with the flipped coin, our uncertainty can be both about what we cannot know, and what we don't know. But how does it make us feel?

How do you respond to uncertainty?

Psychological studies, as well our own experiences, reveal a wide variation in how we all react to the conscious awareness of ignorance that we call uncertainty. Our responses can be broken down into those that are *cognitive* (how we think), *emotional* (how we feel), and *behavioural* (what we do). Table 1.1 lists different ways in which researchers have described possible reactions, and you may want to pause and consider where you lie on these axes: for example, when faced with uncertainty, do you deny it or acknowledge it, does it make you fearful or courageous, do you try to avoid it or approach it? Of course, your response may depend on the context, just as an individual's appetite for risk-taking can vary across different domains;[4] I have known people who seemed to take huge physical risks, and yet were very cautious with their money.

Numerous scales have been developed to measure an individual's intolerance of uncertainty, eliciting responses to statements ranging from 'Unforeseen events upset me greatly' to 'When it's time to act, uncertainty paralyses me.' Those who score highly, and so find it difficult to tolerate uncertainty, may also be at increased risk of clinically significant anxiety and depression.[5]

There are guides for coping with uncertainty, and although this book will

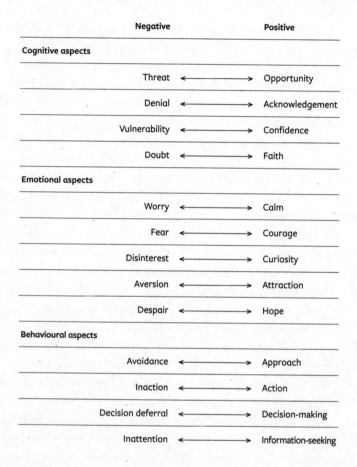

	Negative			Positive
Cognitive aspects				
	Threat	←	→	Opportunity
	Denial	←	→	Acknowledgement
	Vulnerability	←	→	Confidence
	Doubt	←	→	Faith
Emotional aspects				
	Worry	←	→	Calm
	Fear	←	→	Courage
	Disinterest	←	→	Curiosity
	Aversion	←	→	Attraction
	Despair	←	→	Hope
Behavioural aspects				
	Avoidance	←	→	Approach
	Inaction	←	→	Action
	Decision deferral	←	→	Decision-making
	Inattention	←	→	Information-seeking

Table 1.1
Cognitive, emotional and behavioural aspects of responses to uncertainty, showing the potential range of responses.[6]

not be offering self-help advice, I can't resist telling my own story. My father (Cecil's son) had been an enthusiastic traveller, but as he got older he increasingly suffered from what he called 'travel fever' (*Reisefieber* in German, *resfeber* in Swedish), a vivid term for the acute anxiety felt before a journey, essentially due to uncertainty about all the things that could go wrong. This eventually stopped him from going on holiday. So when I started to suffer similar apprehension before travelling, I consulted a psychotherapist. She recommended a small piece of cognitive behavioural therapy (CBT), which involved acknowledging the mental and physical symptoms of anxiety but telling myself that these were essentially indistinguishable from feelings of excitement about the prospect of a journey. 'Thinking' my way from threat to opportunity – as in the first line of Table 1.1 – has been reasonably effective at reframing my response to the considerable uncertainty of travel.

Anticipating an adventure is not the only situation where people might relish uncertainty. When around 1,000 adults in Germany[7] were asked, 'Would you want to know today when you will die?', 88% said no (8% were uncertain, and only 4% said yes). When asked whether they wanted to hear the result of a pre-recorded soccer match, 77% said no, although 23% would want to. And when it came to wanting to know what they were going to get for Christmas, the majority (60%) didn't want to know, with 33% being uncertain and only 7% saying they did. Sometimes we just prefer not to know.

And even if we do want to know, we can still live with, and even welcome, our uncertainty. Theoretical physicist Richard Feynman claimed, 'I'm smart enough to know I'm dumb', and was comfortable with not fully understanding things, saying, 'I can live with doubt and uncertainty and not knowing.' This sets a fine example for how to deal with the inevitable unknowing in our lives.[8]

Not everyone manages to express such humility. As we shall see later, politicians and official bodies may feel they have to exude absolute confidence, particularly when they want to reduce anxiety and reassure the public. After bovine spongiform encephalopathy (BSE), popularly known as 'mad cow disease', was discovered in British cattle, it was uncertain whether it could be passed to humans, but the government maintained that British beef was safe and in 1990 the then Minister of Agriculture was widely publicized eating a beefburger with his four-year-old daughter on a visit to the East Coast Boat Show.* A subsequent inquiry found that the government had been preoccupied with preventing an overreaction to BSE, and this led to them denying uncertainty as to the possible harms.[9] Since then, over 170 people in the UK have died after contracting variant Creutzfeldt–Jakob disease (vCJD) from eating infected beef.

This brings us to the thorny area of risk in its broadest sense, encompassing anything nasty that could happen to either an individual or to society. Psychologists such as Paul Slovic talk about two complementary approaches to such threats, *risk as feeling* and *risk as analysis*; analogous to Kahneman's dual system we met in the Introduction. This book is primarily concerned with an

---

* Neither father nor daughter seems to have suffered ill-effects from their meal, although careful examination of the images shows that the bite-mark in the daughter's burger is not that of a small child.

analytic approach to risk and uncertainty, using numbers, statistical models, and so on, but it's the feelings about risk that tend to dominate our personal attitudes to the perils we may face.

Research by Slovic and others in the 1980s showed that when non-experts were asked about 'risks', their perceptions related more to the characteristics of the possible event, known as the **hazard**, than a reasonable probability of it actually happening. For example, a lion in a strong cage is a hazard, but not a risk, as long as the door is kept shut. Flying in a commercial plane is a hazard, as there is clearly potential to be harmed when five miles up in a rather heavy machine, but is a negligible risk (again provided the door is kept shut). The characteristics that influence risk perception break down into two broad axes, reflecting whether the hazards are 'not-dread/dread' and 'known/unknown'.[10] A hazard is more 'dread' if it is uncontrollable, involuntary, fatal, inequitable and increases risk to future generations – think nuclear accidents. A potential threat is more 'unknown' if it is unobservable, novel, and ill-understood – think attitudes to electromagnetic radiation from mobile-phone masts. Familiar activities like cycling, while potentially risky, are neither unknown nor dread.

Our concerns have changed somewhat over the successive decades – in the 1980s, one of the top threats on the 'unknown' axis was microwave ovens (although I admit I still have suspicions about this mysterious technology). And threats from climate change and AI do not seem to fit naturally along these axes. However, the basic lesson still holds – our concern tends to be not so much related to the uncertainty about whether something will happen, as uncertainty about what it will be like if it does. In the words of H. P. Lovecraft, 'The oldest and strongest emotion of mankind is fear, and the oldest and strongest kind of fear is fear of the unknown.'[11]

Summary

- Uncertainty is a relationship – with a subject who observes, an object that they are uncertain about, a source, a mode of expression, and sometimes an emotional response.
- Broadly, we can have *aleatory* uncertainty about the future, which we cannot know, or *epistemic* uncertainty about the present or past, which we do not know.
- Uncertainty is personal, and our own knowledge can mean we have very different uncertainty to someone else.
- Cognitive, emotional and behavioural responses to uncertainty vary hugely between people, and extreme intolerance to uncertainty can be a source of anxiety and depression.
- There are circumstances where we prefer ignorance.
- We need the humility to acknowledge uncertainty.
- Our personal concern about potential threats tends to be dominated by our uncertainty about what could happen, rather than our uncertainty about whether it will happen.

CHAPTER 2

Putting Uncertainty into Numbers

'We demand rigidly defined areas of doubt and uncertainty!'
—Douglas Adams, *Hitchhiker's Guide to the Galaxy*

We have seen how uncertainty is best thought of as a relationship expressing 'your' ignorance about something tangible. But ignorance is not all or nothing, and when we use expressions such as 'likely' and 'almost certain' in our daily language, we are essentially communicating *degrees of uncertainty*, and a natural next step is to be more precise and put our uncertainty on a numerical scale. This might have helped avoid a disastrous misunderstanding.

―――――

After Fidel Castro's revolutionaries took power in Cuba in 1959, the US Central Intelligence Agency (CIA) plotted with Cuban exiles to overthrow the new regime and restore a US-friendly government. By the time President Kennedy was inaugurated in January 1961, plans were well advanced, but when the US Joint Chiefs of Staff evaluated the proposal for an invasion, they were somewhat sceptical and reckoned it had only around 30% probability of success. When Brigadier General David Gray drafted a report for President Kennedy, he translated this number into 'a fair chance', by which he meant 'not too good'.

But Kennedy apparently interpreted 'a fair chance' as meaning the odds were reasonable, and later gave his support to the invasion.[1] On 17 April 1961, the 1,500 Cuban exiles who landed at the Bay of Pigs on the south coast of Cuba were met with strong resistance, led by Fidel Castro himself – over a hundred were killed and most of the rest captured. The operation was a complete fiasco, a massive embarrassment to the US, and led to Cuba being drawn closer to Russia – the subsequent missile crisis of 1962 came perilously close to a nuclear confrontation.

In his book *Bay of Pigs: The Untold Story*, Peter Wyden reports that it never occurred to Gray that not using a numerical probability might lead to misunderstanding. The Bay of Pigs has also been used as a case study in 'groupthink', when dissenting opinions are silenced. General Taylor, who carried out an inquiry into the disaster, later told Wyden, 'There is a time when you can't advise by innuendos and suggestions. You have to look him in the eye and say, "I think it's a lousy idea, Mr President. The chances of us succeeding are about one in ten." And nobody said that.'

———

If I told you that constipation was a 'common' side effect of taking statins, what proportion of people do you think would suffer this complication if they took the drug? When 120 patients taking statins were asked this question, the average response was 34%.[2] But the true rate is vastly lower, at around 4%. The reason that constipation is officially considered a 'common' side effect is because the European Medicines Agency (EMA) and the UK Medicine and Healthcare products Regulatory Agency (MHRA) dictate that any side effect with an incidence between 1% and 10% is labelled in the patient information leaflet as 'common' – and anything above 10% is 'very common'.[3]

This example reinforces the Bay of Pigs story in showing the dangers of using words to express magnitude, as they can mean very different things to different people. Within the professional health community, the assumption is that side effects are rare, and so even an incidence of 4% is considered common. But this is not how the word is used in ordinary language.

Common, rare, a lot, and so on are vague descriptions of frequency used in everyday language. Even more common are expressions of uncertainty; think how often you say *could, might, maybe, perhaps, likely, possible.* I've argued that it would be better to use probabilities if possible, but since people may be reluctant to put their uncertainty into numbers and wish to use only their familiar linguistic terms, this raises the vital question:

What do we mean by terms like 'likely'?

On 22 January 2010 the UK Terrorism threat level was raised to 'severe', which is officially defined as meaning 'an attack is highly likely'.[4] Given the way most people might interpret 'highly likely', this sounded fairly terrifying, so then Home Secretary Alan Johnson felt obliged to say, 'This means that a terrorist attack is highly likely, but I should stress that there is no intelligence to suggest that an attack is imminent.'[5] Fortunately, there was no terrorist attack.

Numerous studies have shown that the way such words are interpreted can vary substantially between people and contexts. For example, when 5,000 people from 25 countries were asked how they would interpret 'likely' in terms of a percentage probability, the median (middle) response was 60%, but there was huge variation, with one in ten responses lying outside the wide range of 25% to 90%.[6]

This vagueness has naturally led to attempts to standardize the use of such terms, so at least there can be some agreement within specific contexts. One of the most widely used 'translations' was developed by the Intergovernmental Panel on Climate Change (IPCC) and is shown in Table 2.1. Note that the median interpretation of 'likely' by the public (60%) is not even included in the interval (66–100%) mandated by the IPCC, and a general finding is that the public interpretation of these terms is conservative, in the sense of being nearer 50% than the rules shown in the table.[7]

As an example of its use, in 2014 the IPCC reported that 'The period from 1983 to 2012 was *likely* the warmest 30-year period of the last 1,400 years in the Northern Hemisphere', and then at the bottom of the page reminded the reader of the definition that 'likely' means 66–100%. We note the rather wide, and overlapping, intervals in Table 2.1; in fact, since the claim is not described as 'very likely', we might more accurately interpret the claimed probability as lying between 66% and 90%.

Since the debacle at the Bay of Pigs, there have been continued efforts within the intelligence community to be more transparent about their degree of uncertainty. A NATO technical report with the wonderful title 'Variants of vague verbiage' summarized the current use of 'scales of estimative probability' in agencies around the world;[8] Table 2.2 illustrates the differing translations just for the word 'likely'.

This is just a sample of the attempts to standardize communication and, as we shall see in Chapter 9, many agencies also recommend using measures of *analytic confidence*.

Verbal expressions are often preferred by communicators since they can

Term	'Likelihood' of the outcome (probability)
Virtually certain	99–100%
Extremely likely	95–100%
Very likely	90–100%
Likely	66–100%
More likely than not	50–100%
About as likely as not	33–66%
Unlikely	0–33%
Very unlikely	0–10%
Exceptionally unlikely	0–1%

Table 2.1
Probability intervals corresponding to different verbal terms, as mandated by the Intergovernmental Panel on Climate Change (IPCC).[9]

avoid an inappropriately precise probability for the truth of claims, although, perhaps paradoxically, research suggests that such precision means that consumers of scientific claims often prefer to have numbers.[10] The risk of misinterpretation of verbal terms can only increase in audiences with differing first languages. It is now generally recommended that if words are going to be used in formal communications, they should be defined in terms of numerical ranges, and that audiences are repeatedly reminded of the 'translation'. In practice, these are frequently ignored by readers,[11] but this does not reduce their importance.

––––––––

We have seen that the phrase 'fair chance' was misinterpreted before the Bay of Pigs invasion in 1961. Fifty years later, the advice to presidents had become more numerical, and more diverse.

Before the famous raid in 2011, what probability was given to Osama bin Laden being in the compound in Abbottabad?

Agency using 'likely'	Mandated interpretation as a probability range
NATO	60–90%
Canadian Intelligence Assessment Secretariat (also 'probable', 'probably')	70–80%
US Intelligence Community Directive (ICD) 203 (also 'probable', 'probably')	55–80%
UK Defence Intelligence Assessment Probability Yardstick (also 'probable')	55–75%
Norwegian Intelligence Doctrine (also 'probable')	60–90%
Intergovernmental Panel on Climate Change	66–100%
European Food Standards Authority	66–90%

Table 2.2

Examples of different agencies' mandated interpretations of the word 'likely'.

After a ten-year manhunt following the attack on the World Trade Center on 11 September 2001, US Intelligence thought they may have identified Osama bin Laden living in a compound in Abbottabad in Pakistan. But they were not certain, and on 28 April 2011 leading members of the cabinet and other staff met and discussed the options. They had differing opinions, with some advising caution and others recommending that the raid should go ahead. But there were also some numerical assessments available – Barack Obama later said, 'Some of our intelligence officers thought that it was only a 40 or 30% chance* that bin Laden was in the compound. Others thought that it was as high as 80 or 90%. At the conclusion of a fairly lengthy discussion where everybody gave their assessments I said: this is basically 50–50.'[12] Obama apparently left the meeting and said he would let them know his opinion. In the morning he approved the raid.

It is unclear what lay behind Obama's '50–50'. If this was an actual estimate based on a pooling of the opinions being offered, it seems rather low. It

---

* Obama is following the standard practice of using 'chance' as a synonym for 'probability', but in this book I will try to refer to 'chances' only when the numbers are generally agreed from common understanding, rather than based substantially on judgements.

may be just a shorthand for 'we don't know'. Hopefully it does not represent the inappropriate assumption that, if we don't know, then it's 50:50.

We now know that bin Laden was in the compound, and was killed, presumably vindicating those who had given a high probability to his presence there. Some have argued that the wide diversity of views among the intelligence advisers should have been condensed into a single probability assessment before being presented to Obama[13] but, personally, I feel a decision-maker should know when his advisers disagree – Obama needed to synthesize what he was hearing and take ultimate responsibility. Obama is reported as saying, 'In this situation, what you started getting was probabilities that disguised uncertainty as opposed to actually providing you with more useful information.'[14] But I would disagree – far from disguising the uncertainty, the probabilities brought it out into the open, rather than relying on 'vague verbiage'.

––––––––––

The stories in this chapter have, I hope, encouraged the idea that it is better to try to put numbers on our degree of ignorance or, conversely, on our confidence. Some may struggle to do this. Even worse, some people may be deluded about the knowledge they have and claim certainty, or at least high confidence, about facts that are not the case. Fortunately, a simple quiz format with a carefully chosen scoring system shows that it is straightforward to quantify our uncertainty and quickly reveal who is being over-confident.

Do you know what you don't know?

Consider the questions listed below – in each case, either (A) or (B) is the correct answer. The rules are simple:

1. Decide which answer you feel is most likely to be correct.
2. Quantify your confidence on a scale from 5 to 10. So if you are certain (A) is correct, then you should give it 10/10, but if you are only around 70% sure, then it gets 7/10. If you have no idea, then give 5/10 to either choice.
3. No cheating.
4. No cheating.

1. Which is higher?	(A) The Eiffel Tower in Paris (B) The Empire State Building in New York
2. Who is older?	(A) The Prince of Wales (William) (B) The Princess of Wales (Kate)
3. Which is larger?	(A) Croatia (B) Czech Republic
4. Which has the larger population?	(A) Luxembourg (B) Iceland
5. Which has more words?	(A) The Old Testament (King James Version) (B) War and Peace (in English)
6. Which had the highest IMDb rankings (2023)?	(A) Godfather 2 (B) Paddington 2
7. Which is bigger?	(A) Venus (B) Earth
8. Which is furthest north?	(A) New Delhi (B) Kathmandu
9. Which weighs more?	(A) A London double-decker bus (empty) (B) Two average male African elephants
10. Who died first?	(A) Beethoven (B) Napoleon

Try answering these questions using rules 1 to 4 before checking the answers at the end of the chapter. Table 2.3 shows how you should score yourself when the true answer is revealed.

If you are absolutely correct and gave 10/10 to a correct answer, then you score twenty-five on that question. But if you are completely wrong, and gave 10/10 to the incorrect answer, then you lose seventy-five. If your confidence was 5/10 for either answer, then you stay where you were. It is clear that the scoring is *asymmetric*, punishing failure more than rewarding success, so there is a steep penalty for being confident and wrong – it is a very harsh teacher.

This is not arbitrary punishment but a consequence of designing a scoring rule that encourages honesty. It can be shown (see end of chapter) that if, say, you are 70% sure of option (A), then your expected score is maximized if you said 7/10 for (A), rather than exaggerating and claiming 10/10 for (A). Such a scoring rule is called 'proper'.

Your confidence that your answer is correct (out of 10)	5	6	7	8	9	10
Score if you are right	0	9	16	21	24	25
Score if you are wrong	0	-11	-24	-39	-56	-75

Table 2.3
Scoring rule for quiz, after you have given a confidence rating of between 5 and 10 for the answer you think is most likely to be correct. Spot the pattern in the scores.

Have you worked out the pattern in the scores? Try subtracting twenty-five from each of the numbers in Table 2.3. It then becomes clear that the penalties (negative scores) of −1, −4, −9, and so on are dependent on the *square* of the probability given to the wrong answer. This scoring rule is therefore known as a *quadratic* score – it is also a version of what is known as the **Brier score**, after a meteorologist who promoted the rule in the 1950s as a way of training and evaluating weather forecasters when giving probabilities of future events such as rain. At the end of the chapter we see that if we just used the probabilities rather than their squares in the scoring, this would wrongly encourage us to exaggerate rather than be honest.

How did you do, answering the ten questions? When using these quizzes with school audiences, I have found it reveals three broad classes of people; those who have

- A reasonable positive score, say above 80 for ten questions, who know quite a lot.
- A fairly low score, near 0, from people who are cautiously aware of what they don't know, and so tend to give 5, 6 or 7 as answers.
- A large negative score from people who don't know much but *think they do* (in my experience with younger audiences, this trait of over-confidence seems to be more common in male students). You do not want such people as your advisers.

Note that a negative score means you did worse than just answering 5 for every question, which is essentially the strategy of someone, or something, who knows absolutely nothing at all about the answers and just responds '5' to everything. Like a chimpanzee might do.

These are epistemic uncertainties, and the confidence rating can be considered as your personal probability for having chosen the correct answer (when divided by 10 to turn, say, 7 into 0.7 or 70%). This simple quiz therefore has deep lessons. It shows that epistemic uncertainties can be quantified as probabilities, which are necessarily subjective and expressed by an individual on the basis of their available knowledge. By 'subjective' we mean that, although they are numbers, they are not properties of the outside world which can be measured; we measure time with clocks, weight with scales, distance with rulers, but there is no instrument to tell us a probability – it is always a judgement or a calculation based on assumptions. The numbers you gave should not be thought of as embodying some 'true belief' that we could find if we dug deeply enough into your mind, but were *constructed* depending on the context – in this case, a quiz.

But for these judgements to be actually useful, people's probabilities need to have some reasonable properties. First, they should ideally be *calibrated* to the real world, in the sense that if someone gives a probability of 7/10 to a series of events, then around 70% of those events should actually occur. Second, the probabilities should *discriminate*, in that events that occur should be given higher probabilities than those that do not – if we answer 5 to every question we might well end up calibrated (assuming roughly equal numbers of correct answers are A and B), but we are not showing any skill whatsoever. It can be shown that a proper scoring rule rewards both calibration and discrimination,[15] and that good weather forecasters are both calibrated and discriminatory.

By being explicit about the probabilities for events, we can avoid the either/or mentality requiring a simplistic forecast of what is going to happen, which is then shown to be either right or wrong. But it can be difficult to shift people away from this binary way of thinking. Nate Silver of the FiveThirtyEight website had a good reputation for predicting election results, and on 8 November 2016, the day of the US presidential election, he gave a 28.6% probability for Donald Trump to win.[16] This is less than a one in three probability, but Trump, of course, went on to win, and Silver was widely accused of failure because of not calling the election for Trump, even though probabilistic forecasts never 'call' anything.

Silver's probability is roughly equivalent to giving a 7 to response (A) to a question in the quiz and then finding (B) is the right answer. This is hardly a major error,* particularly as Silver's assessment was more pro-Trump than other com-

---

* More imaginatively, we could also consider the election being fought in three parallel worlds, being won by Hillary Clinton in two worlds and by Donald Trump in one. We just happen to inhabit the world where Trump won.

mentators', and Andrew Prokop had written in *Vox*, 'Nate Silver's model gives Trump an unusually high chance of winning. Could he be right?'[17] Of course, Silver was neither right nor wrong, although if he repeatedly gave probabilities such as this, a reasonable scoring rule would identify his poor performance.

———

Since we need to have immediate answers to give feedback, these sorts of quick quizzes necessarily have to assess epistemic uncertainty of facts and historical events. But similar techniques can be used to evaluate forecasters of the future. A panel of experts can be compared according to their scores on test questions, and simple scoring rules enable us to identify people whose opinion is worth taking seriously. Those with high scores may be given additional weight when making group judgements.[18]

This was put to the test in a long series of experiments by a team headed by political scientist Philip Tetlock. Their Good Judgement Project included hundreds of enthusiastic amateurs making forecasts – they were not asked to say what was going to happen, but to give probabilities for tightly defined and verifiable events that would be resolved within a reasonable length of time, such as 'Will Italy restructure or default on its debt by December 2011?' (asked on 9 January 2011).[19] Once the events were known to either have occurred or not, the probabilities were scored using the Brier scoring rule. The synthesized judgements won a major forecasting competition.

When looking at who got the best scores, Tetlock's team found that it made little difference whether the forecaster was conservative or liberal, optimist or pessimist. What mattered was *how* they thought, rather than *what* they thought. So what sort of thinking did best?

Are you a fox or a hedgehog?

I've got to admit I found *War and Peace* rather heavy going, but I do remember the brilliant battle scenes – how they were told from the viewpoint of a single individual who had no idea of what was going on, with no view of a grand plan. Leo Tolstoy excelled at showing characters being buffeted by circumstances and simply trying to make the best of what came their way. But Tolstoy was in deep internal conflict. In contrast to what he depicted so cleverly in his writing, privately he desperately wanted to believe in a grand principle that governed the way the world works. When philosopher Isaiah Berlin wrote his now-famous essay about Tolstoy's dilemma, he called it 'The Hedgehog and the Fox', after a line in a poem by Greek poet Archilochus, 'The fox knows many things, but the hedgehog knows one big thing.' In Berlin's words, Tolstoy was 'a fox bitterly intent upon seeing in the manner of a hedgehog'.

Think of the people you know, either privately or as public figures. Are they hedgehogs, with one overarching view of the world through which they interpret all around them? Or are they foxes, with no big principles or philosophy, who adapt to what turns up, changing their minds along the way? Politicians, of course, tend to be hedgehogs, but some are more pragmatic and foxy than others.

Now, who would you trust most to make predictions about the future: a confident hedgehog or an uncertain fox? In Tetlock's interpretation, hedgehogs have one grand theory, such as being a Marxist, Christian, libertarian, and so on, which they use as a basis for predictions, allowing them to make claims with great confidence. Whereas foxes are sceptical about grand theories, cautious in their forecasts, and are prepared to adjust their ideas when confronted with new evidence. Tetlock found that foxes were much better at predicting than hedgehogs, with hedgehogs being particularly poor at subjects they thought they knew a lot about – they were just too confident (as you may have found when doing the quiz above).

In his book *Future Babble*,[20] journalist and Tetlock-collaborator Dan Gardner identifies three characteristics of good forecasters:

- *Aggregation*: they use multiple sources of information, have an openness to new knowledge and are happy to work in teams.
- *Meta-cognition*: they have an insight into their own thinking, and the biases we all have, such as looking for confirmation of pre-set ideas.
- *Humility*: they have a willingness to acknowledge uncertainty, admit errors and change their minds. Rather than saying what is going to happen, they are prepared to give only probabilities for future events, acknowledging both the known unknowns and the unknown unknowns.

I personally aspire to have these qualities in my judgements, although I struggle to have a sufficiently open mind and awareness of my own fixed ideas. I also look for these features in people who interpret what is happening and prognosticate about what is going to happen. So when someone is telling you what is in store for you, the country or the world, ask yourself – are they a hedgehog or a fox?

Assessing probabilities

The quiz may suggest that getting people to assess probabilities is just a fun game, but it can be a very serious activity. With colleagues, I have interviewed many cancer specialists to elicit their beliefs about the effectiveness of new treatments, which allowed us to assess the overall probability of a planned clinical trial coming up with a convincing result. Not all the doctors were enthusiastic about expressing an opinion and so we sat them down, an interviewer on each side, and would not let them get up until we had extracted a probability distribution for the possible benefit in survival from the new treatment.[21] Such interrogations are now used routinely when planning drug trials in the pharmaceutical industry,[22] ideally conducted face to face, although interactive software exists.

There is no point going through this process unless the subjects are experts in their fields, with extensive relevant experience. They also need training in assessing probabilities, with the rapid feedback which can be provided by the sort of quiz you have just done. This should counter the tendency to be too confident, although care must be taken to avoid known biases – for example, if an event is more salient in someone's mind, because of either anxiety or recent coverage, then it may be thought more likely to occur. Similarly, while it can be useful to start with a ballpark figure, care must be taken that there is not too much 'anchoring' to an initial judgement. There is no best way to elicit probabilities, which points towards interactive questioning using multiple 'frames', such as using both '95% survival' and '5% mortality', or both '10%' and '10 out of 100'.

A traditional approach is in terms of reasonable betting odds; for example, if you are willing to take a three to one bet on an event occurring, it must mean that you assess a probability of more than 25%.* But these thought experiments are mixed up with attitudes to gambling and your sense of the value of money, and so a better approach can be a comparison with a known probability. I own a cardboard 'probability wheel', comprising a yellow circle which I can cover to any desired proportion with a blue overlay. I can then ask someone, which do you think more likely, event X occurring, or a random

---

* Essentially, if you are willing to place a £1 bet at these odds, then you will get back £4 if the event occurs, and so in order to expect (on average) to end up making a profit, you must have more than 25% probability of winning. Technically, if an event is assessed to have probability p, then the odds are $p/(1-p)$, which in this case is $0.25/0.75 = 1/3$, or to 3 to 1 against.

dart landing in the blue rather than the yellow area? The area can then be adjusted until the subject is indifferent.

When we were eliciting whole probability distributions from the doctors we spoke to, we would ask them to assign 100 points to different areas of the scale, essentially letting them build a bar chart. It is well known that people's distributions tend to be too narrow, through excessive anchoring to a central estimate, and so we tried to overcome this overconfidence by asking if they were really, really sure the effect could not be outside their stated range. They were very pleased when we let them get up and go.

Rather than just use a single opinion, we then took a simple average of the clinicians' distributions, although, as previously mentioned, it is possible to upweight those with proven expertise at probability assessment.[23] Betting exchanges, where people are both making and accepting bets, provide another source of group judgements, and allow us to get an idea of what 'the market' thinks are reasonable odds at the moment. For example, we can look at the consensus probability that Barack Obama would win the US presidential election in 2008, as expressed each day* in the year preceding the election. His probability of winning started at only 7%, and then steadily rose to 60% when he got the nomination in June 2008. It briefly dropped to 45% when Obama's opponent John McCain went into the lead following the collapse of Lehman Brothers bank in September, but then Obama's probability rose steadily to 100%. These numbers are not an 'objective' statement about the world, and there is no 'true' probability – they are a reflection of collective subjective judgements given the current state of knowledge. The rapid change in the probabilities after new information shows the participants in the betting exchange had good fox-like behaviour – their collective judgement could be said to display the 'wisdom of crowds'.

The stories in this chapter are intended to convince you that, whenever possible, uncertainty should be put into numbers, which can help both avoid misinterpretation and provide a sound basis for evaluating claims made about uncertain events. Of course, if we have good data relevant to the judgement we are trying to make, then we should use statistical models to help assess probabilities (see Chapter 8).

---

* Using the Intrade betting exchange, which ceased trading in 2013.

Not everyone thinks it is reasonable to express our epistemic uncertainty about past events in terms of probabilities. Members of the Court of Appeal in England clearly stated in 2013 that 'you cannot properly say that there is a 25% chance that something has happened . . . Either it has, or it has not.'[24] I have some sympathy – I would not refer to epistemic probabilities as 'chances'. In any case, Supreme Court judge Lord Leggatt later disagreed, saying it is reasonable to put probabilities on past events provided there is good evidence giving a 'justified belief'.[25]

This reinforces the fact that human judgement is always an integral part of any assessment of uncertainty, and it's only fair to admit that it can be challenging to put this into numbers. So it may be tempting to try to get by with just using words – after all, that's what we do in everyday conversations.

There do seem to be two types of situation when it may be reasonable to claim there is 'deeper' uncertainty that cannot be readily quantified:

1. Although the question and context are clearly defined, the underlying evidence is either poor or may change dramatically, which makes us reluctant to plump for a number or even a range. We shall cover this issue of 'low confidence' in Chapter 9.
2. We just don't know enough about what is going on, and so cannot even list the possible outcomes, let alone put probabilities on them. This is true 'deep uncertainty' (see Chapter 13).

In the meantime we will accept the challenge of using numbers to express uncertainty, which means that we need to understand the underlying theory of probability, or, as it used to be known, 'the Doctrine of Chances'.

Incidentally, you may be wondering whether our doctors were any good at assessing the benefits of new treatments. It turned out that they tended to be over-optimistic, although given their considerable uncertainty, there was generally no strong conflict with the data that was observed later. And in one notable example, when assessing the benefits of a new radiotherapy regimen for lung cancer in 1989, the experts' best judgement was a 24% reduction in monthly mortality risk.[26] When the trial was finally completed and published twelve years later, in 2001, the reported reduction was . . . 24%![27] Very impressive, but there may have been a smattering of luck.

Summary

- Words alone are poor at communicating degrees of uncertainty, since their interpretation can vary hugely between people, languages and contexts.
- There have been numerous attempts to create 'translations' between everyday words and probability ranges.
- We can put numbers on our uncertainty, and scoring rules provide a way of evaluating how good those numbers are.
- Scoring rules reward foxes, and not hedgehogs.
- We can elicit probabilities from people, but they should know a lot about the issue being assessed and the process should be interactive.
- Probabilities can, and should, change rapidly as new information arises.
- Whenever possible, we should use numbers to express our uncertainty.

Answers to quiz

(given in bold)

1. (A) 300m (330m to tip)	vs **(B)** 381m (443m to tip)
2. (A) Born 21/6/82	vs **(B)** Born 9/1/82
3. (A) 56,000km²	vs **(B)** 79,000km²
4. **(A)** 523,000	vs (B) 328,000
5. **(A)** 610,000	vs (B) 590,000
6. **(A)** 9.0	vs (B) 7.8
7. (A) 6,051km radius	vs **(B)** 6,371km radius
8. **(A)** 28.6° N.	vs (B) 27.7° N.
9. **(A)** 12.4 tonnes	vs (B) 10.9 tonnes
10. (A) 1827	vs **(B)** 1821

Why the scoring rule in Table 2.3 rewards honesty

Suppose my honest probability for option B was 70%, and so I chose 7 as my confidence level. Then I feel I have a 70% probability of gaining 16, and a 30% probability of losing 24, and so my 'expected'* score is $(0.7 \times 16) - (0.3 \times 24) = 4$. But suppose I was arrogant and chose to exaggerate and claim 10/10 confidence. Then I have a 70% probability of gaining 25, and a 30% probability of losing 75, and so my expected score is $(0.7 \times 25) - (0.3 \times 75) = -5$, which is lower than if I had chosen to express my true opinion. So although I could be lucky in this instance, on average it will pay me to be honest.

However, suppose we used the rule shown in Table 2.4, which is linear and symmetric.

This rule may superficially seem reasonable, as it essentially penalizes by the distance from the correct answer. Using the previous example, my expected score by being honest is $(0.7 \times 10) - (0.3 \times 10) = 4$, as before. But my expected

---

* 'Expectation', in the technical sense, is covered in the next chapter, but can be thought of as an 'average' score.

score if I exaggerate is $(0.7 \times 25) - (0.3 \times 25) = 10$! So this 'improper' scoring rule encourages people to lie about their uncertainty.

Your confidence that your answer is correct (out of 10)	5	6	7	8	9	10
Score if you are right	0	5	10	15	20	25
Score if you are wrong	0	−5	−10	−15	−20	−25

Table 2.4
An inappropriate scoring rule that encourages exaggerated and dishonest claims.

CHAPTER 3

Taming Chance with Probability

The previous chapter focused on putting numbers on our personal uncertainty about specific events, based on our judgement and unique to ourselves. But if you learned about probability in school or college, you probably followed a very different route, through flipping coins, throwing dice, lotteries and other situations in which 'chance' operates, including notorious questions about socks in drawers.

It's now time to turn to this more traditional approach, historically rooted in games and gambling. It may seem like going back to school, but from simple games we can intuitively grasp the formal rules of probability. This may help you answer some challenging examination questions, and even organize a lottery.

––––––––

A few years ago I asked our local butcher for a leg of lamb, but made an unusual request: that he include the ankle joint. I then did a messy, very amateur, dissection and extracted the small bone embedded in the ankle which provides the articulation between leg and foot – this is known as the *talus* bone in a human, or the *astragalus* in animals, often translated as *knucklebone*. As shown in Figure 3.1, its shape means that it has four possible faces on which it can land when thrown on to a surface. For at least 5,000 years, people from a wide variety of cultures – from Greece to Mongolia – have been gambling on the throws of knucklebones or using them to tell their fortune.

The four possible faces are nowhere near equally likely to come up: I threw mine 200 times and got the percentages listed in the caption to Figure 3.1,

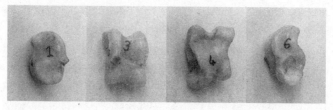

Figure 3.1
A lamb's astragalus, as extracted by me, and the four possible ways in which it can land. Based on 200 throws, I got face 1 in 10% of throws; face 3 in 43%; face 4 in 36%; face 6 in 11%.

while statistical historian Florence Nightingale David reported similar frequencies of approximately 10%, 40%, 40%, and 10%.*

A popular game among the Romans was to throw four astragali and bet on the outcome, with a 'Venus' throw being when the four faces were all different. They were also used as oracles in temples, with a Venus considered favourable, and 'dogs' (all showing face 1) unfavourable.† The great Roman historian Suetonius reports that the poet Propertius once said, 'When I was seeking Venus with favourable tali, the damned dogs always leapt out.'[1]

Gradually the oddly shaped astragalus was replaced by more symmetric dice as the choice for gambling, with a terracotta example from the Indus Valley around 3,000 years ago having numbers 1 to 6 inscribed in the standard pattern in which opposite faces add to 7. But although huge amounts of money must have been staked on the outcomes of small objects being thrown on the ground or a table, nobody seems to have taken a dispassionate analytic approach to what was going on until the 1500s. Which raises the question:

For millennia people were throwing things and gambling on the way they landed. So why did the idea of probability become established only comparatively recently?

---

* An appropriate name for a female statistician, given her namesake's pioneering work in statistical analysis and graphics.

† Assuming the proportions mentioned in Figure 3.1 hold, the probability for a Venus throw is about 4%, and for a 'dog' is 0.01% (1 in 10,000).

There have been many explanations for this odd quirk of history: that classical civilizations valued logical proof rather than experiments; that early gambling devices such as astragali did not have precise symmetries; that the play of chance was considered to be in the hands of the gods; or that numerical systems in the West were inadequate (imagine doing maths with Roman numerals) until around 1200, when Leonardo Pisano, better known as Fibonacci, popularized the Hindu-Arabic system, with base 10 and a zero. For whatever reason, the very idea of numerical chances was just not thought of.

Perhaps we should not be surprised. As already noted, probability is an elusive phenomenon, incapable of direct observation and measurement. Furthermore, as we have seen in Chapter 1, it can have a dual nature: as a measure of epistemic uncertainty expressing a reasonable degree of belief about what could be known, but isn't, and an aleatory measure of future randomness or chance about events that have not been decided yet. And it was through the second path that the breakthrough came, when gamblers finally started to analyse the games they had been staking so much money on.

Gerolamo Cardano lived in Italy from around 1500 to 1571, during the time of rapid artistic and mathematical developments known as the Renaissance. He began life as an illegitimate child disfigured from plague, and by his own admission was 'hot-tempered, single-minded and given to women', but became a wealthy and famous doctor.[2]

He also made, and lost, a lot of money from gambling, and around 1550 decided to put his accumulated wisdom into a book, *Liber de ludo alae* (*The Book of Games of Chance*).[3] In short chapters he covers ways of cheating ('it is of the greatest advantage to have your own supporters if you wish to win unjustly'),[*] the role of luck, and the difference between cards and dice. But his fame rests on being the first recorded person to note that the symmetry of dice means that the numbers should be equally likely to come up, in the apparently innocuous phrase 'I can as easily throw one, three, five as two, four, six.'

By focusing on the *enumeration* of all possibilities, he got tantalizingly close to an idea of probability. A common game was to throw two dice and to bet on the total of the two faces, and he saw that there were 36 possible basic outcomes of the roll of one die and then another, comprising (1,1), (1,2), and

---

[*] Ivey clearly understood this when he was playing at Crockfords.

Second throw

	1	2	3	4	5	6
1	2	3	4	5	6	7
2	3	4	5	6	7	8
3	4	5	6	7	8	9
4	5	6	7	8	9	10
5	6	7	8	9	10	11
6	7	8	9	10	11	12

First throw

Table 3.1
The thirty-six possible basic outcomes of throwing two dice, with the resulting sum of the faces.

so on. He termed the list of all possible outcomes the 'circuit', and these are shown in Table 3.1 with the resulting totals shown in the grid.

He then calculated the number of ways in which, say, a total of 10 could be obtained, as a fraction of the total number of possible outcomes: 'the point 10 consists of (5,5) and (6,4), but the latter can occur in two ways, so that the whole number of ways of obtaining 10 will be $\frac{1}{12}$ of the circuit.' It may seem fairly obvious now, but it was a major step to realize that with two dice there is only one way to throw a total of 2, but many more ways to get, say, a total of 7.

So, although he never explicitly wrote it down, Cardano is generally credited as being the originator of what is known as 'classical' probability, taught in schools all over the world: out of a set of equally likely outcomes, what proportion are 'favourable'? For example, if you want to throw a total of 7, then 6 out of 36 outcomes are 'favourable', and so the probability is $\frac{1}{6}$.

It is important to distinguish the raw *outcomes* of the sequence of throws, which can be assumed equally likely, from any *events* of interest which are calculated from the outcomes, for example the total, the maximum throw, minimum throw, and so on. Each of these is a mapping from a set of basic outcomes to a single number, and the technical term for the resulting event is a **random variable**. Even if the raw outcomes are equally likely, the possible values of the random variable are not equally likely, as Cardano identified when he saw that 3 possible outcomes all mapped to a total of 10.

In this way a **probability distribution** is induced on the random variable and, following Cardano, we can use Table 3.1 to get the probability distribution for the sum of two fair dice, as shown in Figure 3.2.

The average of the total, when weighted by the probabilities of each event, is 7.[*] This is known as the **expectation** of the random variable, also known as its **mean**, and will feature strongly throughout this book.

Note the explicit assumptions that (a) the dice are symmetric and thrown in a way that makes it reasonable to assume that the faces are equally likely to occur, and (b) that they are **independent**, in that the result of the first throw does not influence the second throw. It would become tedious to keep on repeating these caveats, but it should always be kept in mind that any assessed probability is conditional on numerous assumptions.

Cardano did get some things wrong. Although he was admirably clear that his analysis assumed 'the die is honest', unfortunately he forgot this injunction when it came to discussing astragali, when he implicitly assumed that there were equal chances for four faces to appear, and claimed a Venus 'will happen 6 times as often as a throw with all alike'. This was hopelessly incorrect: Figure 3.1 shows the two faces 3 and 4 are strongly favoured, and so, using my observed proportions, the probability of getting all four faces alike is around 5%, compared to 4% for their all being different (the Venus throw). The fact that he had not noticed this suggests he rarely, if ever, played with knucklebones, preferring the more sophisticated (and expensive) dice. He also thought that, for fair dice, 'the chances are equal that a given point will turn up in three throws', meaning, for example, there was a 50:50 chance of throwing a '3' in three throws. The correct answer is 42%, a bit under 50:50.[†]

Cardano lived until he was seventy-four, but, as with many brilliant people, his children were a disappointment; one son was a habitual criminal who became an official torturer and executioner, while, rather ironically,

---

[*] The calculation is $\left(2 \times \frac{1}{36}\right) + \left(3 \times \frac{2}{36}\right) + \left(4 \times \frac{3}{36}\right) + \left(5 \times \frac{4}{36}\right) + \left(6 \times \frac{5}{36}\right) + \left(7 \times \frac{6}{36}\right) + \left(8 \times \frac{5}{36}\right) + \left(9 \times \frac{4}{36}\right) + \left(10 \times \frac{3}{36}\right) + \left(11 \times \frac{2}{36}\right) + \left(12 \times \frac{1}{36}\right) = 7$.

[†] Using the standard rules (listed later in this chapter), the probability of not getting a single 3 in all three throws is $\frac{5}{6} \times \frac{5}{6} \times \frac{5}{6} = 0.58$. So the probability of at least one 3 is $1 - 0.58 = 0.42$. Presumably he thought that since we would expect a 3 to turn up once in six throws, it should have a 50:50 chance of appearing in three throws. But this would only hold if we were *guaranteed* that a 3 would occur in six throws, which would only happen if there could be no repeats.

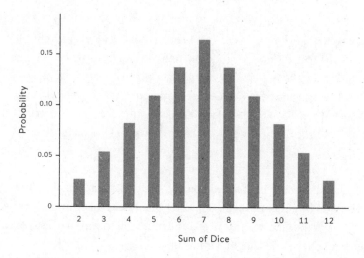

Figure 3.2
The probability distribution for the sum of two dice, assuming perfectly symmetric dice that are fairly and independently thrown.

the other son was executed for poisoning his wife. As an additional misfortune, Cardano's book on games of chance was not published until 1663, ninety years after his death, by which time all his ideas had been attributed to others.

────────

While Cardano's manuscript was lost in his papers, some of the best minds in Europe started to work on games of chance. Galileo Galilei wrote extensively (and tediously) about ways of counting the number of favourable outcomes, while in the mid-1650s Pierre de Fermat and Blaise Pascal had a correspondence that is often credited as the foundation of probability theory, although like Cardano and Galileo, they were essentially concerned with enumerating possible outcomes – the word 'probability' was not used in its modern sense until Jacob Bernoulli's 1713 seminal work *Ars Conjectandi* (*The Art of Conjecturing*).

One of the challenges faced by Fermat and Pascal was the 'problem of points', set 200 years earlier: if a game is interrupted, how should the stake be fairly divided? Below is a (very fictitious) example.

Having nothing better to do, Romeo and Juliet are playing a game in which they flip a (fair) coin, with Romeo winning a stake of 80 ducats if three heads comes up first, and Juliet winning if three tails comes first. The coin has been flipped three times, and the sequence stands at THT. Then dawn breaks, Romeo declares he has to leave, and the game must be abandoned. How should the stake be divided?

Romeo says it should be divided equally, with 40 ducats each, as the game could be won by either of them. But Juliet points out that this does not seem fair – she is in the lead and just needs one more tail to win. We need to calculate how much the odds are in her favour.

One approach is to list all the possible ways the game could develop. By five flips it must be over, as there must be at least three heads or three tails, and so if we suppose they continue until the bitter end, the game could end as

- **THT** HH – Romeo wins
- **THT** HT – Juliet wins
- **THT** TT – Juliet wins
- **THT** TH – Juliet wins

If we assume the coin is fair, then all these 'possible futures' are equally likely, and so there is a three out of four chance that Juliet wins.

This does seem a somewhat clumsy method, since the final two games could have been stopped after four flips. Instead, we might think of what we would expect to happen if, in some bizarre Groundhog Day scenario, they played the same game 100 times. This is shown in Figure 3.3, in a format known as an *expected frequency tree*.

We would expect Juliet to win on the next flip in 50 of the continued games, and in 25 after two more flips, one head and one tail. We don't know which of these equally likely 'possible futures' will occur, and therefore a reasonable probability for Juliet eventually winning a particular game would again be $^{75}/_{100} = \frac{3}{4}$.

An alternative approach is to draw a *probability tree* as in Figure 3.4, which shows how the different possible futures may arise, with an assumed probability of going down each split of a branch. There is a probability $\frac{1}{2}$ that their

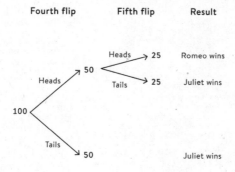

Figure 3.3
An expected frequency tree showing what we would expect to happen were Romeo and Juliet's interrupted game to continue 100 times. Juliet wins in 75 out of 100 possible futures.

next flip (the fourth) is a head, which would put them in equal position, and in this case then there is a further probability $\frac{1}{2}$ that either could win on the fifth flip. In order to match the frequencies in Figure 3.3, we need to obtain a final probability of Romeo winning of $\frac{25}{100} = \frac{1}{4}$, which we simply obtain by multiplying the probabilities down the branch to get $\frac{1}{2} \times \frac{1}{2} = \frac{1}{4}$. Similarly, to match the frequencies of Juliet winning in Figure 3.4 ($\frac{75}{100} = \frac{3}{4}$), we add up over the ends of the relevant branches to get $\frac{1}{4} + \frac{1}{2} = \frac{3}{4}$.

We have shown three alternative ways of solving the 'problem of points': (a) enumerate all possible future games, (b) look at what we would expect to happen in many repetitions, and (c) calculate the probabilities of a single continued game. In each case we obtain the probability of Romeo or Juliet winning is in the ratio 25:75. But that does not necessarily tell us how the stake should be divided.

Remember, the expectation of a random variable is the average outcome, weighted by the relevant probabilities. When the game is interrupted, Romeo has a 25% probability of winning 80 ducats, and a 75% probability of winning 0, so his expected win is $(80 \times 25\%) + (0 \times 75\%) = 20$, while Juliet has an expected win of 60 ducats. Pascal, Fermat and everyone else working on the problem of points implicitly assumed that it was fair to divide the stake according to the expected winnings of each contestant, and so Romeo should receive 20 and Juliet 60.

The issue of deciding the outcome of a game that has been interrupted may seem rather niche, but it is of huge importance in major one-day cricket

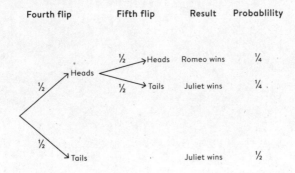

Figure 3.4
A probability tree for a single instance of Romeo and Juliet's inter-rupted game. Juliet has $1/2 + 1/4 = 3/4$ probability of winning.

tournaments. When play is interrupted by rain or bad light, an assessment is needed for how many runs to set as a target for the opposing team to beat. The Duckworth–Lewis–Stern (DLS) method, originally invented by statisti-cians Frank Duckworth* and Tony Lewis, sets this target as an estimate of the expected number of runs given the available 'resources', in terms of wick-ets and balls left to play. The method has been repeatedly revised, becoming even more complex, and its implications are not always clear. For example, in a one-day match between England and the West Indies in 2009 in Guyana, the West Indies coach withdrew his team due to bad light, believing that they had met the Duckworth–Lewis–Stern target. But he had not taken into account that a player had been out with the last ball, and this unfortunate miscalculation meant that England were then declared the winners by one run. Which must be a lesson to everyone facing decisions made by formulas.

———

Although the probability tree in Figure 3.3 for Romeo and Juliet's game may seem intuitive and rather basic, it manages to reveal the basic rules of probability:

---

* I knew Frank, who sadly died just as this book was being completed – he was a charming cricket enthusi-ast who was bemused by becoming internationally famous in his retirement from being a statistician in the nuclear industry.

1. *The probability of an event is a number between 0 and 1*: impossible events are given probability 0 (for example, that someone other than Romeo or Juliet wins), and certain events given probability 1 (that someone wins).

2. '*Complement rule*': the probability of an event not happening is 1 minus the probability of its happening. For example, the probability of Juliet winning is 1 minus the probability of Romeo winning: $1 - \frac{1}{4} = \frac{3}{4}$.

3. *The addition, or the OR rule*: add probabilities of mutually exclusive events (meaning they cannot both happen at the same time) to get the total probability. For example, the probability of Juliet winning is $\frac{3}{4}$, since it can occur through 'a tail at the fourth throw' with probability $\frac{1}{2}$, OR a 'head + tail' with a probability of $\frac{1}{4}$.

4. *The multiplication, or the AND rule*: multiply probabilities to get the overall probability of a sequence of independent events (meaning one does not affect the other) occurring. For example, the probability of a head AND a head is $\frac{1}{2} \times \frac{1}{2} = \frac{1}{4}$.

These rules mean that, after drawing a probability tree

- To get the overall probability of reaching the end of a branch, multiply the probabilities of the splits in the branch (Rule 4).
- To get the overall probability for an event (say Juliet winning), add up the overall probabilities for each of the branches that lead to that event (Rule 3).

And that's it! The whole of probability theory can be summarized by these simple ideas. They will turn out to be remarkably useful.

————

In many games of chance, it is reasonable to assume that repeated observations are independent – the result of the first coin flip does not affect the probabilities linked to a second flip. But often we need probabilities that change depending on previous results – this is known as **conditional probability**. For example, as every card-counter knows, once an ace has been dealt from a pack of cards, the probability that the next card is an ace goes down.

So, as warned, here come the socks.

I am too lazy to match up clean socks. One morning there are two purple socks and four green socks mixed up in a drawer, and I take two out at random without looking at them. What is a reasonable probability that I have a matching pair?

This is known as *sampling without replacement*, as I keep each sock that I pick rather than putting them back in the drawer, and so the probabilities for a subsequent sock are changed. We can construct a similar table to that for Cardano's two dice (Table 3.2), but noting that some cells of the table are impossible, since the same sock cannot be drawn twice. Simple enumeration gives a probability of $\frac{7}{15}$ for a pair.

We can also construct a probability tree (Figure 3.5) for the possible outcomes. For example, there is a $\frac{2}{6} = \frac{1}{3}$ probability that the first sock is purple; there are then five socks left, of which one is purple, so the probability that the second is also purple is $\frac{1}{5}$. Rule 4 says that, to obtain the overall probability of each type of pair (purple/purple, purple/green, green/purple, green/green) we should multiply the probabilities down the branches, giving the values shown on the right of Figure 3.5. So the total probability of getting purple + purple is the probability of the first sock being purple, multiplied by the conditional probability of the second being purple, given the first was purple.[*] This is $\frac{2}{6} \times \frac{1}{5} = \frac{2}{30}$.

The probability of getting a pair is, by Rule 3, the probability of purple + purple plus the probability of green + green, which is $\frac{2}{30} + \frac{12}{30} = \frac{14}{30} = \frac{7}{15}$, reassuringly the same answer as obtained through exhaustive enumeration of all possibilities.

These may seem fairly basic ideas, but as we shall see, they caused a lot of consternation when they appeared in a school examination question.

––––––

On the bright, sunny morning of Thursday, 4 June 2015, over 100,000 15- and 16-year-old students in England were sitting nervously in their examination

––––––

[*] In notation, this can be written as Pr(both socks purple) = Pr(first purple) × Pr(second purple | first purple), where 'Pr( )' should be read as 'probability for', and the '|' symbol should be read as 'given' or 'conditional on'. This is a special case of the general rule for the joint probability for two events A and B; Pr(A and B) = Pr($A|B$) × Pr(B).

		Second sock				
	P1	**P2**	**G1**	**G2**	**G3**	**G4**
P1		Pair				
P2	Pair					
G1				Pair	Pair	Pair
G2			Pair		Pair	Pair
G3			Pair	Pair		Pair
G4			Pair	Pair	Pair	

(First sock labels the rows on the left.)

Table 3.2

All possible outcomes of drawing two socks at random from two purple (labelled P1 and P2) and four green (G1 to G4). Out of 30 possible outcomes, assumed equally likely, 14 lead to a pair of socks. So a reasonable probability for drawing a pair is $^{14}/_{30} = ^{7}/_{15} = 47\%$.

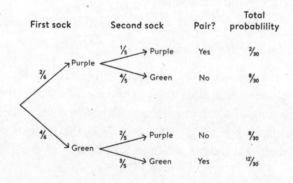

Figure 3.5

A probability tree for picking two socks at random from a drawer with two purple and four green socks. The probability of a matching pair is $^{2}/_{30} + ^{12}/_{30} = ^{14}/_{30}$.

halls, about to start the Edexcel GCSE Higher Tier Mathematics examination.[4] They turned over the paper, and after the usual geometry and algebra they got to question 19, where they found this question about Hannah's sweets.

There are n sweets in a bag. 6 of the sweets are orange. The rest of the sweets are yellow. Hannah takes a random sweet from the bag. She eats the sweet. Hannah then takes at random another sweet from the bag. She eats the sweet. The probability that Hannah eats two orange sweets is $1/3$.

(a) Show that $n^2 - n - 90 = 0$. [3 marks]
(b) Solve $n^2 - n - 90 = 0$ to find the value of n. [3 marks]

Many students, and their families, found the question not just difficult, but baffling. Comments on social media included 'You wottttttt, where did that even come from?!!' and 'Why, Hannah, why do you have to have them sweets?'[5] A petition to make Edexcel change their grade boundaries was signed by thousands, and Hannah's sweets became a national talking-point, with solutions being offered on TV and news media. So pretend that you are sixteen years old and back in an examination hall – can you answer the problem of Hannah's sweets? Please try before reading the solution below, which uses the level of algebra expected of the school students.

Solution to Hannah's sweets

There are n sweets, 6 of which are orange, and so the probability that the first sweet is orange is $6/n$. After Hannah has eaten that sweet, there are $n - 1$ sweets remaining, of which 5 are orange. Therefore the probability that the second sweet is orange is $5/(n-1)$. The probability that both are orange is the product of these probabilities, which we are told is equal to $1/3$. So

$$\frac{1}{3} = \frac{6}{n} \times \frac{5}{(n-1)}$$

Rearranging gives

$$n \times (n-1) = 90$$

which means that

$$n^2 - n - 90 = 0$$

as required for part (a). For part (b), either solve this quadratic equation using the standard formula (which was provided on the examination paper), or see that the equation factorizes into

$$(n - 10)(n + 9) = 0.$$

For this to be true, n must be either 10 or -9, and since n must be positive, we conclude there were $n = 10$ sweets in the bag; 6 orange and 4 yellow.

So that's six marks towards your GCSE Mathematics!

———

All these examples are based on enumeration of equally likely outcomes, and some of these enumerations can get complicated. Unfortunately, this has meant that ways of calculating the number of 'permutations and combinations' has tended to become a major, and generally unpopular, part of teaching probability, in spite of these counting techniques having nothing to do with uncertainty. However, a certain familiarity is useful, and so I rather apologetically include some details.

As a simple example, suppose I have forgotten the four-digit PIN for my new bank card, although I remember it contains the digits 6, 7, 8 and 9, in some order. I am standing in front of the cash machine/ATM. How many tries would I need in order to be certain to enter the right code?[*]

This is essentially another 'sampling without replacement' problem. The first digit I enter can be any one of four options; the second can be any one of the three remaining; the third can be one of the two remaining; and I have no choice about the final digit. This means there are

$$4 \times 3 \times 2 \times 1 = 24$$

different numbers that I can enter – these are the permutations of 6, 7, 8, 9, such as 6789, 6879, 6897, and so on. There is some useful mathematical notation for the total number of permutations: 4!, known as '4 factorial' – at school we called it '4-shriek'. The general rule follows that, if we have n things, then there are $n! = n \times (n - 1) \times (n - 2) \times \ldots 1$ orders in which they can be placed; a useful, but apparently odd, convention, is to assign 0! the value 1.

As the next example shows, n-shrieks can get remarkably big.

> Take a pack of cards and shuffle them well. Has anyone ever, in the whole of history, produced exactly the same order of cards after a good shuffle?

---

[*] Although in practice you are likely to be locked out after a few tries.

Your shuffle has produced a particular order of the fifty-two cards in a pack. From the general rule about permutations, there are

$$52 \times 51 \times \ldots \times 1 = 52!$$

different orders. If you start putting this into a calculator, it will rapidly grow to a very large number indeed, roughly 8×10^{67} – that's 8 with 67 zeros after it. This is a bit more than the number of atoms in our galaxy, the Milky Way.[6]

It is estimated that around 100 billion people have ever existed on the earth, and suppose we very generously assume that on average they lived seventy years (their biblical sell-by date of three-score years and ten). If all of them had done nothing in their lives except make one shuffle every ten seconds, this would only make 2×10^{19} shuffled packs. This is clearly a massive overestimate of the actual number of shuffles there have been in history, but even so there would be only a 1 in 10^{48} chance of a match with your personal shuffle.

So we can be extremely confident that your exact shuffle has never been done before. Stephen Fry on the television programme QI categorically claimed that nobody had ever made the shuffle that he had just done,[7] but we cannot be so absolutely, totally, logically certain. There is a far higher probability that in the whole of history there have been two identical shuffles, but this is still practically zero.*

———————

You may have sensed my lack of enthusiasm for teaching methods of counting rather than the important ideas of probability. But it is undeniable that calculating how many ways there are to arrange things has played a fundamental role in the development of probability and statistical science, as well as being vital when designing a lottery.

This next question dates from the 1700s and forms part of a major intellectual development that influences vast swathes of statistical modelling today.

———————————————————————————————

* Assuming the ludicrous number of 2×10^{19} shuffles in history, this makes 2×10^{38} possible pairings of shuffles. This would mean there is less than a $2 \times 10^{38} / 8 \times 10^{67} \approx 1$ in 10^{29} chance of two good shuffles, in the whole history of card-playing, ever having been the same.

If I flip a fair coin many times, what is the probability of getting any particular number of heads?

If we assume that each flip is independent and a head occurs with probability $\frac{1}{2}$, then each particular sequence is equally likely to occur; for example, with four flips, HHHH is just as likely as HTTH, and each has probability $\frac{1}{2} \times \frac{1}{2} \times \frac{1}{2} \times \frac{1}{2} = \frac{1}{2^4} = \frac{1}{16}$. This may not seem intuitive, as the second sequence may feel more 'typical' than one in which only heads occur, just as a lottery draw of ball numbers (27, 22, 6, 48, 50, 7) may seem more likely than (1, 2, 3, 4, 5, 6). But our intuition would be, as is often the case with probability, wrong.

However, if we are just counting the *number* of heads as a random variable, rather than the actual sequence, then the possible events are no longer equally likely. It is more likely that we get two heads and two tails than all heads – this time our intuition would be correct. But how much more likely?

Just as for Cardano throwing two dice, if we want the distribution of this random variable, we need to work out the number of ways of getting a particular total number of heads. Fortunately this is provided by what has come to be known as 'Pascal's triangle', even though Pascal acknowledged he did not invent it. Incidentally, this is a fine example of *Stigler's Law of Eponymy*, which states that anything named after someone was not, in fact, invented by that person; naturally, historian of statistics Stephen Stigler admits that he did not make up this law.[8]

Figure 3.6 shows the first few rows of the triangle, which has a satisfying pattern, with each entry being the sum of the two entries above it.

For example, if we flip four coins one at a time, Row 4 shows there are sixteen possible sequences, with

- 1 way to get 0 heads (TTTT)
- 4 ways to get 1 head (HTTT, THTT, TTHT, TTTH)
- 6 ways to get 2 heads (HHTT, HTHT, HTTH, THHT, THTH, TTHH)
- 4 ways to get 3 heads (HHHT, HHTH, HTHH, THHH)
- 1 way to get 4 heads (HHHH)

Since each particular sequence is equally likely, this means it is six times as likely to get two heads as it is to get no heads.

Row								
0					1			
1				1		1		
2			1		2		1	
3		1		3		3		1
4	1		4		6		4	1
5	1		5	10		10	5	1
6	1	6	15	20		15	6	1

Figure 3.6
Pascal's triangle. Row *n* shows the number of ways to get a particular number of events in a sequence of *n* opportunities.

Some notation now becomes useful, and apologies if it looks complicated. The rth element in the nth row of the triangle is the number of ways of choosing r distinct elements from a pool of n, and follows a basic formula written as $_nC_r$ or $\binom{n}{r}$, where $\binom{n}{r} = \frac{n!}{r!\,(n-r)!}$.* For example, if we have 12 children standing against the wall bars in a gym lesson, and we want to choose a five-a-side team, there are $_{12}C_5 = \binom{12}{5} = \frac{12!}{5!\,7!} = 792$ different teams that could be chosen.

The number of heads in a sequence of coin flips follows what is known as a **binomial distribution**,† and examples are shown in Figure 3.7. The heights of the bars are proportional to the entries in Pascal's triangle, which are known as binomial coefficients.

These distributions allow us to assess the probability for different deviations from an exactly even split between heads and tails, for example, for 100 flips, there is a 6% probability of getting a split of 60:40, or more extreme, in favour of heads or tails.

These calculations can get very clumsy, but fortunately, nearly 300 years ago, mathematician Abraham de Moivre realized that, for larger n, there is a nice smooth approximation to the binomial distribution, as illustrated by the gentle curve for 1,000 flips in Figure 3.7(d). We first need to introduce

---

* If we have n flips, the number of sequences with precisely r heads, and therefore $(n-r)$ tails, is $_nC_r$. For example, $\binom{4}{2} = \frac{4!}{2!(4-2)!} = \frac{4 \times 3 \times 2 \times 1}{2 \times 2} = \frac{24}{4} = 6$.

† A Binomial distribution for coin flips shows the probability of getting exactly r heads in n flips is $\binom{n}{r}\frac{1}{2^n} = \frac{n!}{r!(n-r)!}\frac{1}{2^n}$.

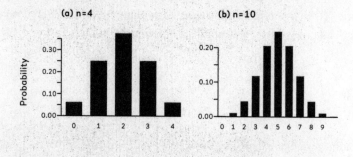

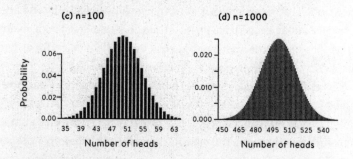

Figure 3.7
Binomial probability distributions for the number of heads in *n* flips of a fair coin. As *n* gets large, this distribution tends to a smooth, approximately normal curve.

the idea of a **variance** of a distribution, which is one summary of its spread.* The smooth approximation found by de Moivre is now known as a **normal** or **Gaussian distribution**, and has the same expectation (mean) and variance as the binomial.

Incidentally, from the formula for the normal distribution we find that the probability of getting exactly the same number of heads and tails is approximately $\sqrt{2/n\pi}$.† For example, in 100 flips, the probability of getting exactly 50 tails and 50 heads is approximately $\sqrt{1/50\pi} = 0.08$. It is perhaps curious that π,

---

* The variance of a distribution is the expected size of the square of the deviation of a random variable from its mean. The **standard deviation** is the square root of the variance, so is a sort of average distance of each observation from the mean. The distribution of the number of heads in *n* flips of a fair coin has mean $n/2$ and variance $n/4$.

† A normal random variable with mean m and variance v has a probability density $f(x) = \frac{1}{\sqrt{2\pi v}}\exp\left[-\frac{1}{2}(x-m)^2/v\right]$. If $x = m$ (that is, we observe the expected number of heads), and the variance is $n/4$, then substitution gives $f(m) = \sqrt{2/n\pi}$.

the ratio of the circumference of a circle to its diameter, is intimately connected to the chance of getting equal numbers of heads and tails.

How did Casanova's mathematical skills lead to a successful lottery?

Giacomo Casanova is notorious as a lover, gambler and adventurer, but less known are his formidable skills in mathematics and probability. In *Casanova's Lottery*, Stephen Stigler describes how when Casanova found himself in Paris in 1757 after his famous escape from prison in Venice, he exploited both his analytic talents and his powers of persuasion to head up a collaboration for a state-run national lottery, intended to pay for the École Militaire.[9] The lottery was based on a 'wheel of fortune' containing balls numbered 1 to 90, from which five balls were to be drawn at random – this would now be known as a 5/90 lottery. Punters could buy bets on particular choices of either one, two or three numbers to come up, for a fixed payout if they won. Casanova made it clear that the government would not make a profit at every drawing, but claimed they were guaranteed to win in the long run at the payouts he suggested.

This would have been a reckless proposal if Casanova had not had a good idea of what the chances of making a profit would be. Fortunately, by the mid-1700s there were established techniques for calculating chances of winning lottery draws, using the formulae for combinations described above, and Casanova's skills were sufficient to calculate the chances shown in Table 3.3, and recommend payouts that should make a profit.[10]

Suppose you bought a ticket specifying two numbers, say 20 and 42. There are $_{90}C_2$ = 4,005 possible two-number tickets that you could have bought.* Of these, the winners will occur in the five chosen balls, so there are $_5C_2$ = 10 winning pairs. So $^{10}/4,005$, or 1 in 400.5, of two-number tickets will win, and that is the chance that you chose one of the winning pairs.

Casanova chose his payouts carefully. For example, of three-number tickets costing, say, one livre each, we would expect 1 in 11,748 to win, and that ticket will win 5,200 livres, which means that on average only 44% of the money

---

* From the general combinatorial formula, $_{90}C_2 = \frac{90!}{(88! \times 2!)} = \frac{(90 \times 89)}{2} = 4,005$. Alternatively, we know there are 90 options for the first choice, and 89 for the second. But this means every pair of numbers will be represented twice, for example (29, 62) and (62, 29), and so we need to divide by 2 to get to $\frac{(90 \times 89)}{2} = 4,005$ unique two-number tickets.

Type of bet placed: r	Number of possible tickets: $_{90}C_r$	Number of winning tickets: $_5C_r$	Chance of chosen numbers being in the 5 selected	Payout for a winning ticket	Payout compared to chance
One number (extrait)	90	5	1 in 18	15	$\frac{15}{18} = 83\%$
Two numbers (ambe)	4,005	10	1 in 400.5	270	$\frac{270}{400.5} = 67\%$
Three numbers (terne)	117,480	10	1 in 11,748	5200	$\frac{5,200}{11,748} = 44\%$

Table 3.3
Casanova's lottery, in which a choice is made of either one, two or three numbers from 1 to 90. Five balls are drawn, and the ticket wins if they include the chosen numbers.

staked on three-number tickets will be paid out in prizes. The lottery is therefore bound to make a profit in the long run, although it is not guaranteed to do so for each draw. Overall the lottery paid out 72% of its takings, which is a lot more than the current UK lotteries, which pay out around half of the money staked.[11]

Casanova's lottery was very successful, earning huge amounts for the government – at one point providing 4% of the national income. With relatively small changes it ran from 1758 until 1836: the draw was disturbed neither by the storming of the Bastille in 1789 nor the execution of Louis XVI in 1791, although there was a temporary hiatus when all lotteries were banned for three years during the period of revolutionary terror. It was a convincing demonstration of the importance of careful assessment of probabilities.

Casanova continued his itinerant life of philandering, gambling and failed business ventures, and with his lurid memoirs achieved perpetual notoriety, although sadly not for his mathematical abilities.

———

This chapter has dealt only with situations where we can assume equally likely outcomes, and assess the probabilities of events by simple enumeration of numbers of possible 'favourable' outcomes. But this is very restrictive, and by the early 1700s Swiss mathematician Jacob Bernoulli had given his name to the

Bernoulli trial, which is a random variable that takes on the value 1 if an event occurs, and 0 if it does not, with probability p; for example p would be $\frac{1}{6}$ if we were throwing dice and only interested in whether a six came up. If p is not $\frac{1}{2}$, then all possible sequences are no longer equally likely, and the total number of occurrences of the event of interest is given by a binomial distribution for general p.[*] This formula enables us to calculate, for example, the probability of getting exactly two sixes in twelve dice throws (which comes to 0.3).[†]

Bernoulli also developed the famous *Law of Large Numbers*, which says that if we observe an increasing number of independent Bernoulli trials, the proportion in which the event occurs tends towards p. For example, if 30% of the population believe that the earth is flat, then if we take a big enough random sample and ask their opinion, then the observed proportion of flat-earthers in our sample will be close to 30%. That assumes, of course, that we are not systematically biassing our sample by, for example, interviewing people at a flat-earth convention.

As the sample size increases, the observed proportion will wobble around a lot before it gets increasingly close to 30%. The *gambler's fallacy* supposes there is some magical process by which any initial imbalance is evened out, a classic example being when claims are made about a particular lottery number, or colour on a roulette wheel, being 'due' as it has not occurred for a while. In fact, it is best to think of initial imbalances being *swamped* rather than *corrected*.

Probability theory can tell us how close we should be to the unknown true p, and from this we can construct estimates and intervals for p. Statistical science is therefore a way of taking the uncertainty implicit in sampling, and using probability theory to produce inferences about underlying states of the world.[‡] This is a great achievement, all based on the extraordinary work of some brilliant individuals analysing games of chance more than 300 years ago.

The theory of probability has made further strides, including the basic rules being put on a more rigorous footing by Russian mathematician Andrey Kolmogorov in the 1930s. But while this may tidy up the mathematics, it does nothing to answer the more basic question

What is probability anyway?!

---

[*] A binomial distribution for the number R of successes in n trials, each with probability p, is $\Pr(R=r) = \binom{n}{r}p^r(1-p)^{n-r} = \frac{n!}{r!(n-r)!}p^r(1-p)^{n-r}$.

[†] Using the binomial formula, the probability is $\Pr(R=2) = \binom{12}{2}\left(\frac{1}{6}\right)^2\left(\frac{5}{6}\right)^{10} = \frac{12 \times 11}{2}\frac{5^{10}}{6^{12}} = 0.30$.

[‡] See my earlier book *The Art of Statistics* for a lot more on this!

Warning, the rest of this chapter gets a bit philosophical, but try to hang in there.

The traditional view, taught to generations of students, treats probability as an *objective* feature of the world, imposing some regularity on apparently unpredictable events. Suggestions for its meaning have included

- **Classical probability** based on symmetries, as we've seen in our examples of coins, dice or lotteries, which allows games of chance to be analysed by enumeration of equally likely events. But this is a circular definition, as it requires a judgement of 'equally likely'.
- **Frequentist probability**, which is the theoretical proportion of events that would be seen in infinitely many repetitions of essentially identical situations. This approach is widely adopted in supposedly 'objective' scientific studies. When applied to a specific situation, each single event has to be placed in a *reference class* of events assumed to have the same probability; this may be clear in repetitive contexts such as roulette or lotteries, but in general this choice of the reference class is inevitably a judgement, even if this is rarely made explicit.
- **'Propensity'**, which is the idea that there is some true underlying tendency for a specific event to occur in a particular context, such as my having a heart attack in the next ten years. While this rather mystical, and unverifiable, idea in principle allows unique probabilities to be considered objective, it cannot apply to epistemic probabilities.
- **Logical probability** is the objective degree to which a set of preconditions logically implies a conclusion, and so could in principle lead, say, to a justified 50:50 belief that a flipped and covered coin is heads. But this only applies to very restricted situations.

Note that all these apparently 'objective' interpretations of probability require substantial judgements to actually assign numerical values.

A fundamentally different perspective, which we have taken from the start of this book, is that probability is a *subjective* quantification of personal uncertainty, or what has been called 'partial belief'. But this still leaves us with the challenge of defining what we mean when we say, for example, that the probability that Donald Trump will win the 2024 US presidential election is judged to be around 40%, based on the betting exchanges at the time of writing (which is December 2023 – I have deliberately chosen an example for which readers far enough in the future will know the truth).

The first definitions are all to do with 'rational' decision-making.

- **Indifference with betting on an event with a 'known' probability:** Rather like classical probability, we allow ourselves the idea of 'equally likely', in this case a random number generator producing a number equally likely to be anywhere between 0 and 1. We then can check whether we would be indifferent between betting on Trump being the next President, or the generator coming up with a number of 0.4 or less. This is the electronic equivalent of the cardboard probability wheel mentioned in Chapter 2.

- **Reasonable betting odds:** In 1926, Frank Ramsey[*] showed that all the laws of probability could be derived from expressed preferences for specific gambles. Outcomes are assigned 'utilities', and the value of a gamble is summarized by its expected utility, where the weights in the expectation are governed by the subjective numbers expressing partial belief, that is, our personal probabilities. So, while taking into account our personal valuation of money (Chapter 15), our probabilities are determined by the odds we are willing to accept in a gamble on the 2024 election.[12]

    Ramsey's betting-odds probabilities were not arbitrary. He expected them to be calibrated, in the sense we saw in Chapter 2, so that of all events to which he assigns a probability of 0.4, he expects 40% to occur; he wrote, 'given a habit of a certain form, we can praise or blame it accordingly as the degree of belief it produces is near or far from the actual proportion in which the habit leads to truth'.

- **Maximizing expected 'score':** If we imagine the presidential result being a question in a forecasting competition, where your probabilities are to be assessed using a rule such as the one in Chapter 2, then you would

---

* Frank Ramsey argued that, if your probabilities did not obey the 'laws', then someone could make a series of bets against you that were guaranteed to lose you money – a so-called Dutch book. Ramsey is also the person from history whom I would most like to meet. He was a genius, whose work is still considered fundamental in probability, mathematics and economics, and was Ludwig Wittgenstein's Ph.D. supervisor, translator and friend. He was a big man, weighing 17 stone (238lbs; 108kg), only worked in the mornings and played tennis in the afternoons, drank and enjoyed exuberant parties, had a wife and lovers, and laughed 'like a hippopotamus'. He died in 1930 aged just twenty- six, possibly from Weil's disease after swimming in the River Cam (something I have done regularly). This was a massive loss to British intellectual history – if he had survived, he would have been Alan Turing's supervisor at King's College Cambridge, and no doubt would have helped break the Nazi codes and shorten the war even further than did Turing and the others at Bletchley Park. I made a BBC Radio 4 *Great Lives* documentary on Ramsey, which featured Cheryl Misak, author of the excellent biography *Frank Ramsey: A Sheer Excess of Powers*.

assign a probability of 0.4 to maximize your expected score. In general, to maximize your expected score, your probabilities need to obey the correct laws.

Other interpretations might be considered 'subjective frequentist', in that although they are personal judgements, they represent expected proportions of repeated occasions.

- **Expected proportion of similar situations:** physicist Richard Feynman defined probability as 'our judgement of the most likely proportion of occasions in which an event occurs',[13] which is explicitly subjective and refers to a sequence of similar events. Alan Turing used a similar idea: 'The probability of an event on certain evidence is the proportion of cases in which that event may be expected to happen given that evidence.'[14] Like the frequentist interpretation of probability, these definitions require the current judgement to be embedded in some larger class, but this could presumably be taken to be all situations in which a probability of 0.4 or 40% is given. So Feynman and Turing are essentially saying they expect their probabilities to be calibrated in a series of judgements.
- **Expected proportion of 'possible futures':** Instead of embedding a specific probability in a set of repeated assessments, we might stretch our imagination and think of what might happen were the current circumstances repeated again and again. So we would judge, for example, that Donald Trump would be elected in 40% of the 'possible futures' that could develop from December 2023 to December 2024. I personally find this a useful, although admittedly metaphorical, concept, and it could even appeal to the 'many-worlds interpretation' discussed later.

Of course, some of these subjective 'partial beliefs' will have a stronger justification than others – if I have examined a coin carefully before it is flipped, and it lands on a hard surface and bounces chaotically, I will feel more justified with my 50:50 judgement than if some shady character pulls out a coin and gives it a desultory few turns before catching it. So we can have greater confidence about some judgements, as we shall see in Chapter 9.

It can come as a surprise to those who have learned the mathematics of probability that there is still no agreement as to what it actually is. Or whether, as we now consider, it even exists.

When I was studying mathematics at university in the 1970s, my mentor Adrian Smith* was translating Bruno de Finetti's *Theory of Probability*[15] from the original Italian. De Finetti had developed the ideas of subjective probability in the 1930s,† entirely independently of Ramsey, and began his book with the provocative statement

Probability does not exist.

This may seem quite extreme and, although it is beyond my philosophical pay grade to start discussing what it means for something to 'exist', I interpret this as de Finetti simply proclaiming that probability was not an objective property of the world. I fully absorbed this sentiment in my youth, and in fifty years I have never shifted from the view that probabilities are subjective judgements, even if they are based on considerations of physical symmetries, data analyses or complex models. The only possible exception is at the subatomic quantum level, where it could be argued that truly objective and determined probabilities exist (see Chapter 6).

This means that probabilities are fundamentally different from other numbers that we use routinely, such as those to denote time, distance and temperature. As we mentioned in Chapter 2, vast intellectual resources have been put into measurement with clocks, rulers, thermometers and other instruments, so that the resulting numbers are agreed to be adequate descriptions of the external world, up to whatever accuracy is required. But where is the 'probability-ometer' that enables us to measure a probability? There isn't one, except perhaps in the highly restricted theoretical case of infinitely repeatable identical trials. Indeed, probability might be considered a *virtual* quantity.

But if we accept that probabilities are constructed on the basis of personal judgement, does this mean that any old numbers are OK, provided they obey the rules laid out previously? Can I just say there is a 99.9% probability that I can fly off my roof? Well, I could, but I would soon be proved to be a poor probability assessor if I tried. And this is where the objective, outside world

---

* Now (2024) Professor Sir Adrian Smith, President of the Royal Society, the highest accolade for a scientist in the UK.

† Although his political opinions changed during his life, at that time de Finetti was an enthusiastic supporter of Mussolini's style of fascism; somewhat in contrast to Ramsey's staunch socialism.

comes in – in the *evaluation* of probabilities when tested against reality, say using a proper scoring rule. Or an ambulance.

Fortunately, as we shall explore further in Chapter 6, in practice we don't have to decide whether objective 'chances' really exist in the everyday, non-quantum world – we can take the pragmatic approach of simply acting *as if* they do. Rather ironically, the most persuasive argument for acting as if chances exist was provided by de Finetti himself, in his 1931 work on 'exchangeability'.[16] A sequence of events is judged to be **exchangeable** if our beliefs about each sequence are unaffected by the order of the observations; for example, when assessing the probabilities of whether a geyser is going to erupt on each of a set of days, the actual dates are irrelevant and the intended observations could be in any order. De Finetti brilliantly proved that, if we make this assumption of exchangeability, it is mathematically equivalent to acting *as if* the events on each day were independent, each with some true underlying chance of erupting, and our uncertainty about that unknown chance is expressed by a subjective epistemic distribution. This is remarkable, and rather beautiful – from a purely subjective expression of beliefs, it shows that we should act as if events were driven by objective chances.

The theory of probability underlies all of statistical science and much of scientific and economic activity, and it is extraordinary that such an important body of work has arisen from something that, it can be argued, simply does not exist.

Summary

- People have been gambling for millennia, but rather remarkably the idea of probability was not properly developed until the 1600s.
- The rules of probability can be intuitively derived from considering what we would expect to happen in a large set of repetitions.
- Conditional probabilities, that change as the situation changes, arise naturally when we are sampling without replacement.
- If we can assume a physical process that generates equally likely outcomes, then assessing probabilities becomes a matter of counting 'successful' outcomes that lead to the event of interest.
- Formulae and approximations for counting 'successful' outcomes allow us to establish probabilities of success in games of chance, lotteries, and so on.
- Probability has been interpreted in many ways, both as an objective property of the world and as a subjective judgement.
- If we accept that probability is different from other measures that are routinely used, and is constructed on the basis of subjective judgement, then we can still use the objective world to evaluate the quality of those probabilities.
- However, objective 'chances' may hold at the subatomic quantum level, and it can be useful to act as if they exist in everyday life.

CHAPTER 4

Surprises and Coincidences

In Chapter 1, I defined uncertainty as the 'conscious awareness of ignorance', and the quiz in Chapter 2 and the probability exercises in Chapter 3 have demonstrated how we might put into numbers our conscious uncertainty about events. But for most of our lives we are not so conscious – we toddle along with some rough plan, not thinking about all the possibilities that might happen, adjusting our behaviour to follow an intended path like a lone sailor automatically making small adjustments to keep on course.

But every so often we are shaken out of our complacency by a surprising event. This may be an unexpected misfortune, like that sailor being hit by a freak wave, but it could be a benign and fortuitous concurrence of events – often labelled as a coincidence – that brings our uncertainty into consciousness. Accidents or disasters may shock or harm us, but most coincidences make us smile – they could be considered an 'upside' of uncertainty. I am fascinated by coincidences, even though they almost never happen to me. So a few years ago, when I was making a programme about chance for the BBC, our team in Cambridge started the 'Cambridge Coincidence Collection', which ended up containing about 5,000 stories submitted by the public.[1]

And so to Ron Biederman's trousers.[2] Someone, whom we shall call Doug, reported getting all his clothes stolen at a Miami backpacker hostel and being kindly given a striped shirt from Israel by someone called Ron Biederman. A few years later, Doug was staying in another hostel in London and started talking about Israel to a girl sitting opposite him in the café. The first coincidence they noticed was that they had both met Ron Biederman, and the second was that Doug was wearing Ron's shirt at the time. The girl then stood up to reveal she was wearing the matching trousers, given to her on a kibbutz by Ron Biederman.

To be honest, I thought this was a bit far-fetched, until I got an email asking me about the story – from Ron Biederman himself! He confirmed he had given these clothes away, sent me a photo of him and the trousers, and was so pleased the recipients had met. A very satisfying connection.

————

People seem to like talking about the coincidences they have experienced, but what *is* a coincidence? In a classic academic paper, statisticians Persi Diaconis and Frederick Mosteller used the definition:

> Coincidence: *'a surprising concurrence of events, perceived as meaningfully related, with no apparent causal connection'.*[3]

This contains three necessary elements:

1. The event involves an unexpected connection.
2. It leaps out of everyday circumstances and grabs our attention – we may remember it for the rest of our lives.
3. There is no immediate explanation for why it happened – although, as we shall see later, there have been many theories as to why they happen.

Among the most common themes in our Cambridge coincidence collection are

*   Finding a link with someone you meet: such as two strangers talking in a hotel in Rome, finding they both had sons working in the same company, phoning them up and finding they were sitting at adjacent desks.
*   Meeting someone you know in an unlikely place: such as Mick Preston on holiday in the Pyrenees setting off to the post office with a postcard for his friend Alan, and on the way meeting Alan.
*   An object reappearing: such as being on holiday in Portugal and finding a coat-hanger that belonged to your brother forty years previously.[4]

Some stories are less easily classified, such as the couple who found they had both been born in the same village in Germany, which had only one little hospital and one bed where all the babies were born. So they reckoned they had both been born in the same bed.[5]

All these bizarre events clamour for an answer to the obvious question: 'What's the chance of that?!' Unfortunately most coincidences are not amenable to formal analysis, but we can try to put numbers on some of them. And one of these happened to me.

———

As I said, I rarely experience coincidences. I am too unobservant, and so I would never, as other people have reported, note seeing the same person repeatedly while walking round London. Also, being typically English, I tend not to talk to strangers unless I have been introduced, so I could have sat for hours next to my long-lost twin on a train and would never have realized it. In fact, the biggest coincidence in my life (so far) happened in 2018 while recording a contribution to a BBC radio programme which, rather remarkably, was about coincidences.[6]

I was telling a coincidence story which involved a birthday, which happened to be 27 January. There was a pause, and the interviewer said, '*So David, while you've been telling me that really interesting story about the birthday on 27 January, producer Kate has just been speaking in my ear, telling me that not only is her birthday 27 January, but the engineer that she is working with in the studio right now recording this interview also has a birthday on 27 January.*' Well, what's the chance of that?

A reasonable probability for Kate having this birthday was $\frac{1}{365}$, and assuming the producer was not her twin, the probability of both being born on 27 January is $\frac{1}{365} \times \frac{1}{365}$, which is around 1 in 133,000. So it was really quite surprising, although perhaps not as strange as some of the stories in this chapter. And, rather satisfyingly, it was caught on the recording and features in the programme, a rare instance of a coincidence being captured as it happened.

The same calculation applies to a story that crops up fairly often in the media (presumably supplied by an agency on a weak news day) about a family with three children who are different ages but who share the same birthday.* If we assume birthdays occur randomly throughout the year, then the oldest child 'sets' the birthday, and the probability that their two siblings are born on the same day is again $\frac{1}{365} \times \frac{1}{365} \approx \frac{1}{133,000}$, just as for the extraordinary birthday match during my radio recording. Sometimes the media get this wrong,[7]

———

* Just search on 'three children born on same day' and see all the stories come up.

usually by including the first birthday in the calculation and multiplying by $1/365$ to get 1 in 48 million (if the chances really were this small, perhaps they should wonder why the story appears so frequently).

I was contacted by *More or Less*, the popular BBC radio programme about statistics,[8] to answer a similar question posed by a listener, David, who had been born on 6 February, as had two of his three children.

> In a family of two parents and three children (no twins), what is a reasonable probability for a parent having the same birthday as two of their children?

At first sight it might appear that the probability would again be $1/365 \times 1/365 \approx 1/133,000$. But David has three children and so there are $_3C_2 = 3$ possible pairs that could match, and there is also a choice of two parents, and so I came to the conclusion that my probability would be around $6/133,000 \approx$ 1 in 22,000. Since there are around 1 million families in the UK with three children under eighteen, I could tell David that his was among around $1,000,000/22,000 \approx 45$ similar families in the country. So they were unusual, but certainly not unique.

All these calculations assume that each day has an equal probability of being a birthday. This is not true. First, the family may plan to have children at a particular time of year.* Second, there are fewer births on public holidays, and more around forty weeks after the Christmas holidays – 27 September is the most common birthday in the year. But these deviations from pure chance are not big enough to have much influence on the calculations, and in any case make a match even more likely.[9]

These examples, which I admit are not of world-shaking importance, illustrate a standard technique for analysing apparently unusual events:

A. Assess, if possible, a probability for the specific instance being looked at.

---

* I've been told personal stories of meticulous family planning leading to the children all having birthdays within a week of each other, which at least saves money by sharing parties.

B. Estimate the total number of opportunities for a similar event to happen in a defined context over a particular period.

C. Multiply the answers to (A) and (B) to get an expected number of events.

D. Use this expectation to work out how 'surprising' it is to hear about such an event.

These techniques can be used for both clusters of catastrophes and, as we now see, touching family histories.

———

Some stories take a lifetime to unfold, as shown in the extraordinary memorial to Mr and Mrs Huntrodds in Whitby, Yorkshire,[10] displayed in Figure 4.1.

As you can see, they were both born on 19 September 1600, married on 19 September, had twelve children, and then died within five hours of each other on their joint eightieth birthday on 19 September 1680. Now that's an impressive match.

How unusual would the Huntrodds' shared birthday be today?[11] There are around 13 million cohabiting couples in England and Wales[12] so, if birthdays play no part in people getting together, by chance we would expect around $13{,}000{,}000/365 \approx 36{,}000$ couples to share a birthday. Around 9% of the couples who married in 2001 were the same age. So, again assuming random pairing,

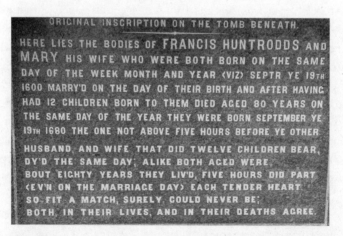

Figure 4.1
The memorial to Francis and Mary Huntrodds at the Church of St Mary, Whitby, Yorkshire, born, married and died on 19 September.

over 3,000 couples would be able to share a single birthday cake with the same number of candles.* A notable example was Joyce and Ron Pulsford from Pagham, West Sussex, who were both born on 8 August 1928 and celebrated their joint eightieth birthday on 08/08/08.[13]

You may not consider the Huntrodds to be a coincidence. After all, they presumably chose to marry, and when to marry, and so the really odd thing is for both of them to die on their joint birthday. It is often reported that birthdays are higher risk, although the statistics can be influenced both by babies who sadly died immediately after birth, and registration errors in which the day of birth is mistakenly copied as the day of death. But was there plague in Whitby in 1680? Did they have an accident at their birthday party? It would be so good to know. They must have been local characters, having so many children, the same birthday, and being so old for the period. They deserve their memorial.

There have always been theories for why coincidences happen, and many have suggested there is some external force that brings about these 'surprising concurrences'. Paul Kammerer developed the idea of *seriality*, claiming that, 'side by side with the causality of classic physics, there exists a second basic principle in the universe which tends towards unity; a force of attraction comparable to universal gravity'. Similarly, psychiatrist and psychoanalyst Carl Jung proposed the existence of *synchronicity*, an 'acausal connecting principle' that explains not only physical coincidences but also premonitions. A similar idea of *morphic resonance* comes from parapsychology researcher Rupert Sheldrake, who has suggested that 'morphogenetic fields work by imposing patterns on otherwise random or indeterminate patterns of activity',[14] and which can explain claimed phenomena such as the feeling that you are being stared at, dogs knowing when their owners are coming home, and so on.

I am afraid I am sceptical about these theories of some external force. I would argue that the undoubted occurrence of extraordinarily surprising events can usually be put down to three main causes:

- The Law of Truly Large Numbers:[15] if there is a sufficiently large number of opportunities, even very rare events will eventually happen.

---

* Thanks to Timandra Harkness for this image.

- Being selective: only remembering the surprising concurrences, and ignoring all the irrelevant predictions, dreams and premonitions that did not happen. This is demonstrated by the plethora of 'psychic' animals predicting results of sporting competitions.
- Contriving the story to make the event more surprising: for example, in tests of extrasensory perception, being generous in declaring a 'match' between two separated people's drawings.

Perhaps the most extraordinary aspect of coincidences is how *few* are reported. For every one that is identified, there must be a huge number of near misses that are not noticed: maybe I *have* sat next to my long-lost twin separated at birth. These 'latent' coincidences must be happening to us all the time, if we only knew the possible connections with the people we encounter.

———————

A classic test of the Law of Truly Large Numbers involves simians and a certain famous playwright.

Will a group of monkeys on typewriters eventually tap out the *Complete Works of Shakespeare*?

For a 2010 BBC Horizon programme about infinity,[16] I installed a Monkey Simulator program[17] and left it running in my office for days.[18] After 113 million imaginary keystrokes (equivalent to around 26 days for 50 monkeys typing one character a second) the best that the virtual monkeys managed to produce was the nine characters 'we lover', which appeared in *Love's Labour's Lost*, Act 2, Scene 1, in a speech by Boyet: 'With that which we lovers entitle affected.'

The *Complete Works* has about 5 million characters and, even ignoring upper and lower case and punctuation, we calculated that each time the monkey starts typing, there is a chance of 1 in $10^{7,500,000}$ of completing Shakespeare. Since $10^{7,500,000} \approx 2^{25,000,000}$, this is about the same chance as flipping a fair coin 25 million times and its coming up heads every time, or winning the lottery every week for 20,000 years. Unlikely, but not logically impossible, and so perhaps worth a try. So with £2,000 of Arts Council funding, in 2003 researchers installed a keyboard for four weeks at Paignton Zoo in a cage with six macaques – Elmo, Gum, Heather, Holly, Mistletoe and Rowan.

Unfortunately, they produced just five pages of text between them, primarily filled with the letter S, and then soiled the keyboard.*

The Law of Truly Large Numbers states that coincidences happen because there are lots of opportunities, and this can give rise to surprises occurring surprisingly often. One of the classic examples, even sometimes referred to as a 'probability paradox', says that, in a group of 23 random people, at least half the time there would be at least one pair with the same birthday. This means, for example, that in over half of football games, two people on the pitch (out of the 22 players and the referee) will share a birthday.

Coincidentally, World Cup squads comprise precisely 23 players, so out of the 32 teams in the 2023 Women's World Cup, we would expect 16 to have players with matching birthdays. There turned out to be . . . 17! Two Nigerian players, appropriately named Glory Ogbonna and Christy Ucheibe, both being born on Christmas Day.[19] In fact we might have expected even more pairings, as elite sports players tend to be at the older end of a year-group.

This is another example of a 'matching problem', in that we have a group of people or other things, and we want the probability of at least one pair sharing a particular characteristic. If we want to calculate the probability of a match, the first lesson is that it is always best to calculate the probability of *no match*, where all the people are different, and subtract it from 1.

To get this probability we can either do a rather complex exact calculation, or a neat shortcut. First the exact calculation. Imagine 23 people in a line, and we want the probability they all have different birthdays. The first person's birthday can be anything; the second birthday must be different from the first, and this has probability $364/365$;† the third birthday must be different from the first and second, and this has probability $363/365$, and so on – this is essentially an example of sampling without replacement, and so the conditional probabilities change. The probability that all 23 birthdays are different is therefore equal to

$$\frac{364}{365} \times \frac{363}{365} \times \ldots \times \frac{343}{365} = 0.997 \times 0.995 \times \ldots \times 0.940 = 0.49.$$

* 'It was a hopeless failure in terms of science, but that's not really the point,' said Geoff Cox, of Plymouth University's MediaLab.
† I ignore leap years throughout this chapter, which makes a negligible difference.

Each of these 22 numbers is near 1, reflecting that each particular individual in the line is likely to have an 'unused' birthday. But when lots of these numbers are multiplied together, the product ends up less than a half – this numerical phenomenon is the source of the unintuitive result.

As promised, there is an alternative, extremely useful, shortcut for getting at such probabilities, which could enable you to amaze your friends and possibly win money off them.

Rule: Suppose we are in a context where there are many opportunities for a particular type of rare event to occur. Then if we expect, on average, m rare events, the chance that none of them will occur is e^{-m}.

Here, e is the exponential constant $2.718\ldots$, obtained as the limit of $\left(1 + \frac{1}{n}\right)^n$ as n goes to infinity. This is an extremely useful number, first discovered (or invented, depending on your philosophy of mathematics) by Jacob Bernoulli in 1683 when working on compound interest, and forms the basis for the idea of **exponential growth**, which substantially raised its profile during the Covid pandemic.* The rule above can be derived directly† from the definition of e.

Let's go back to the birthday problem and see a more direct way of showing that there is a more than 50% chance of a match in 23 people. If we take any pair of people in a group of 23, there is a $\frac{1}{365}$ chance they have a matching birthday. But there are many possible pairings, in fact $_{23}C_2 = \frac{(23 \times 22)}{2} = 253$ – this is the number of handshakes required if each person was told to shake hands with everyone else. There are therefore 253 opportunities for a match, each with probability 1/365, and so the expected number of matches is $\frac{253}{365} = 0.693$.‡ Using the rule above, a calculator reveals that the approximate probability of no matches is $e^{-0.693} = 0.499$, and so this simple approximation gets the correct answer that there is slightly more than 50% chance that two will share a birthday.

---

* Compound interest, say in which a deposit of £100 increases at 3% per year, is an example of an initial quantity being multiplied by a fixed amount k in each unit of time. If we define $r = \log k$ (natural logarithms to base e), then this implies $k = e^r$, and after n units of time our initial quantity has increased by a factor e^m. Which is why it is called exponential growth.

† Note that e^{-x} is defined as the limit of $\left(1 - \frac{x}{n}\right)^n$ as n gets large. If each of n rare events has a small probability p of occurring, we expect $m = np$ to happen, and the chance that none occurs is $(1 - p)^n = \left(1 - \frac{np}{n}\right)^n = \left(1 - \frac{m}{n}\right)^n \approx e^{-m}$.

‡ The events are not independent, but the expectation of the sum of a set of Bernoulli trials is the sum of their individual probabilities, regardless of any dependence.

Expected number of events (m)	Approximate probability that no events occur: e^{-m}	Approximate probability that exactly one event occurs: $m\,e^{-m}$	Approximate probability that at least **one** event occurs: $1 - e^{-m}$
0.693	50%	35%	50%
1	37%	37%	63%
2	14%	27%	86%
3	5%	15%	95%
4	2%	7%	98%

Table 4.1
Suppose there are many opportunities for rare events to occur, and a known expected number m. The columns show the approximate probabilities for no events, one event and at least one event. The probability for one event comes from the Poisson approximation discussed below, and numbers are subject to rounding.

Table 4.1 shows some examples of the simple rule for specific values of expected numbers of events, which can be used to solve many sorts of matching problems.

> Suppose a sports team all give their locker keys for safekeeping to the referee, who then drops them in a heap. The referee then hands the keys back at random, and each of the team tries the key they've been given. What's a reasonable probability that at least one player actually gets to open their locker?*

This may seem unanswerable, as I have not said how large the team is. But the crucial observation is that, regardless of the number of players, the expected number of keys returned to their rightful owner is 1† – essentially, as the number of players increases, the chance of each individual player

---

* This is traditionally known as the 'hat-check' problem, involving people checking in their hats at the opera and the person on the counter getting the tickets mixed up. But that seems a little old-fashioned.

† The reasoning goes as follows. Suppose there are n players and therefore n keys. Any specific player has got a $1/n$ chance of getting each key, and so a $1/n$ chance of getting their own, which means that the expected number

getting their own key goes down, but there are more players, and so the total expected number of matches stays the same. Therefore, from Table 4.1, the probability that nobody gets the right key is approximately $e^{-1} = \frac{1}{e} = 0.37$ or 37%, and so the probability at least one player can open their locker is 63%. The approximation is very accurate provided there are at least five members in the team.

All this has been known for 300 years, since French mathematician Pierre Raymond de Montmort analysed the game of Treize in the 1700s. This was a form of Snap in which each of two players shuffled a full suit of thirteen cards, with, say, one having Hearts and one Spades, and they then simultaneously turned their cards over one at a time, claiming a match if they both turned over the same numbered card, say the five of Hearts and the five of Spades. Montmort's methods were later refined by famed mathematicians Nicolas Bernoulli (nephew of Jacob) and Leonhard Euler[20] for different numbers of cards in play; they showed that the probability of a match very rapidly approached $1 - e^{-1} = 0.6321\ldots$: for example, with each player having just five cards, the probability of a match is 0.63. This simple game always favours the person who bets on a match occurring.

––––––––

If you were that sort of person, how could you use these ideas to take money off people? First, you might want to play Treize or Snap assuming winner-takes-all: if you always bet on there being a match, you will win 63% of the time regardless of the number of cards in play.

Your opponent may catch on to this rather quickly, and so here are some other tricks. Diaconis and Mosteller give a simple approximation* for the number of people necessary to be confident there is a close match, to some degree, in birthdays.[21] Table 4.2 shows how this can be used to judge how many people are necessary to have either a 50% or 95% chance of a birthday match of up to three days' difference. For example, if we are prepared to declare a match if birthdays are just one day away from each other, Table 4.2 reveals we only need 13 people

––––––––

of correct keys for each player is $\frac{1}{n}$. Since the expectation of the sum of a set of random variables is the sum of their individual expectations, the total expected number of correct keys for the whole team is $n \times \frac{1}{n} = 1$.

* With n people, the number of pairs is approximately $\frac{n^2}{2}$, and so if the probability that any particular pair matches is $\frac{1}{K}$, then the expected number of matches is around $\frac{n^2}{(2K)}$ and the probability of no matches is $e^{-n^2/(2K)}$. If we equate this to 0.50 and 0.95 and solve for n, we get the required approximations of $1.2\sqrt{K}$ and $2.5\sqrt{K}$.

Gap between birthdays	Probability of 2 random people 'matching': 1 in K	Number needed for probability of match to be approximately 50%: $1.2\sqrt{K}$	Number needed for probability of match to be approximately 95%: $2.5\sqrt{K}$
Same day	1 in 365	23	48
+/- 1 day	1 in 122	13	28
+/- 2 days	1 in 71	10	21
+/- 3 days	1 in 52	9	18

Table 4.2
Approximate numbers of people required to have 50% or 95% probability of a specified degree of matching between birthdays. Suppose a particular match between two people has a chance $1/_K$, then for a 50% chance of a match we need around $1.2\sqrt{K}$ people, and for a 95% chance of a match we need around $2.5\sqrt{K}$ people.

to have an approximate 50% chance of a match.* Once again we see that in 23 people there will be 50% probability of an exact matching birthday, and there is also at least 95% probability that two of them will have birthdays between 2 days of each other, say 6 and 8 June (it would be almost exactly 95% for 21 people, and so even higher for 23). So you are almost certain to win that bet, although it would not seem so impressive.

Another way to baffle and amaze your friends is to ask for the last two digits of their phone numbers and see if there are two that match. Table 4.3 shows that, for example, in a group of 15 people, the expected number of matches is 1.05,† and so the approximate probability of at least one match is 65%, close to the exact probability of 67%.

I have played this game with groups of twenty people, in which I ask them to choose a random number between 1 and 100, and I win if two of them choose the same number. If they genuinely choose at random, I have an 87% probability of winning, and it can be very impressive when matches keep on occurring.‡

---

* The exact probability, assuming random birthdays and ignoring 29 February, is 0.483.

† $15 \times 14/_2 = 105$ pairings, each with $1/_{100}$ chance of a match.

‡ In general, if you have n people around you, get them to choose a number at random between 1 and $n^2/_4$; for example, if you have 30 people, choose a number between 1 and $900/_4 = 225$. The expected number of matches is the number of pairs of people (approximately $n^2/_2$) times the probability of a particular pair choosing the same

Number of people	Expected number of matches between last two digits of their phone numbers (m)	Approximate probability of at least one match between the last two digits of their phone numbers $1 - e^{-m}$	Exact probability of at least one match between the last two digits of their phone numbers
2	0.01	1%	1%
5	0.1	10%	10%
10	0.45	36%	37%
15	1.05	65%	67%
20	1.90	85%	87%
25	3.00	95%	96%
30	4.35	99%	99%

Table 4.3
Approximate and exact probabilities of matches between the final two digits of people's phone numbers. With twenty people, there is an 87% probability of a match.

But people find it extremely difficult to choose random numbers and tend to choose favourites such as 7 and 99, which massively increases my chance of winning but makes the trick less amazing. Phone numbers will be more random, but they can only be played once. Or, on long, boring car journeys with children, get them to record the final two digits on registration plates they see, and bet them their pocket money that in the next twenty cars they will find repeated numbers.* You should have an 87% probability of winning, and also teaching them a valuable lesson about betting.

Entrez, Baron Poisson!

So far we have just looked at the probability of no events occurring, but we may also be interested in the probability of exactly one, two or more. If we assume there are n independent opportunities for an event to happen, each

---

number ($4/n^2$), which is 2. So the probability of no matches is around $e^{-2} = 0.13$. So, if they really pick at random, there is around an 87% probability at least two of the numbers will be the same.

* This idea is stolen from Marcus du Sautoy.

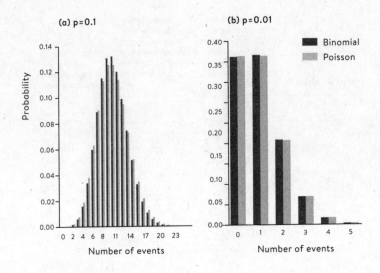

Figure 4.2
Binomial distributions for *n* = 100 and *p* = 0.1 (10%) and
p = 0.01 (1%), compared to Poisson distributions with means 10 and 1.

with probability *p*, then we can use the binomial distribution introduced in Chapter 3. Consider asking a sample of a hundred people about a particular characteristic – say whether they like Marmite.* Suppose the true share in the population is 10% or 1%, and we make the bold assumption that the survey is perfectly carried out, the numbers of respondents liking Marmite will follow the binomial distributions shown in black in Figure 4.2.

In 1711, Abraham de Moivre showed that with large *n* and small *p*, the binomial probabilities could be well approximated by a simpler form, later called a **Poisson distribution**, after Baron Siméon Denis Poisson's formal derivation in 1837.† The Poisson distribution is completely determined by its expectation *m*, which in these cases is *m* = *np* = 10 or 1, and applies to situations where there are a large number of opportunities for a rare event to happen, such as the number of homicides each day in England and Wales,[22] the number of Prussian officers kicked to death each year by their horses, and, as we shall see in Chapter 11, the number of goals in a football match. The distribution allows

---

* For non-British readers, Marmite is a powerfully flavoured spread made from yeast extract that tends to arouse polarized feelings of either exuberant appreciation or extreme disgust.
† Yet another example of Stigler's Law of Eponymy.

us to estimate the probabilities for any number of events occurring,* and so can be used to answer the following type of coincidence, which is rather less innocuous than the previous examples.

> How unusual is it to have three major plane crashes over an eight-day period?

Back in 2014, Malaysia Airlines Flight 17 was shot down over Ukraine on 17 July, TransAsia Flight 222 flew into buildings in Taiwan on 23 July, and Air Algérie flight 5017 stalled and crashed in Mali on 24 July. How surprising is such a tragic cluster?

The extraordinary PlaneCrashInfo[23] website reports that 91 commercial flights containing 18 or more passengers had crashed in the previous 10 years (2004–13), a rate of one every 40 days on average.†

Consider any specific window of 8 days. If planes crash in an entirely unpredictable way at a rate of 91 over 10 years (3,650 days), then we would expect $8 \times {}^{91}\!/_{3,650} = 0.2$ crashes in any particular 8-day window. Using the Poisson distribution with this mean, a reasonable probability that there are at least three crashes would be around 1 in 1,000. So it is very surprising indeed for there to be three crashes between 17 July and 24 July 2014.

But this is not the right question to ask. There is nothing special about these particular eight days, and we are only interested in this specific period because of the crashes. Rather, we should be concerned with whether such a cluster is surprising over some longer interval, say ten years. A rather complex 'scan-statistic' adjustment, that allows for all possible eight-day windows over this period, puts the probability up to 0.59.[24] So there is around a six in ten chance that we should see such a large cluster over a ten-year period, and so

---

* If a Poisson distribution has expectation m, then the probability of observing r events is $e^{-m}\frac{m^r}{r!}$. Remembering the convention that $0! = 1$, the probability of no events ($r = 0$) is therefore e^{-m}, and the probability of one event ($r = 1$) is $e^{-m}m$, as used in Table 4.1.

† This website should be avoided by those who are nervous about flying, as it is full of fascinating detail about plane accidents, including the crash caused in 2010 by the panic after a live crocodile escaped from a passenger's hand luggage. See the next chapter for another story about escaped wildlife on public transport.

it is not at all surprising that such a cluster should occasionally happen. And, reassuringly, the rate of major fatal plane crashes is steadily declining.

––––––––

Statistical analysis rapidly demolished any idea of a common cause behind the plane crashes in 2014, but other clusters of tragic events can lead to suspicions of malicious behaviour. There have been a series of court cases in which a series of deaths or serious events has been linked to a specific individual, leading to suspicions of 'medical murder'. Sometimes some basic statistical analysis reveals clearly that the pattern of events could not be explained by chance alone, as in the case of Dr Harold Shipman, the British family doctor who was eventually found to have murdered at least 215 of his patients over a 20-year period – as I describe in *The Art of Statistics*, he could have been identified as being highly unusual after around 40 deaths, if only someone had been looking at the data.

Shipman was an extreme example of a true serial murderer, but other cases show the caution required before a court should conclude that the events are malicious. Lucia de Berk was a paediatric nurse in the Netherlands who was convicted in 2004 of murdering seven children, and attempting to murder three more. Once she came under suspicion, intense scrutiny identified a series of adverse events in patients under her care. At her trial, a claim was made that, by chance alone, there was only a 1 in 342 million chance of so many deaths occurring while she was on duty.

Senior statisticians re-examined the evidence and concluded a more reasonable probability could be as high as 1 in 25 and, after further medical evidence came to light, de Berk was retried in 2010 and was freed. A report from the Royal Statistical Society later argued that it was essential that professional statisticians should be involved in critiquing any such claim of 'too unlikely to just be a coincidence'.[25] We shall look at similar miscarriages of justice in Chapter 10.

––––––––

Throughout this book I have continually emphasized that any probability assessments are conditional on assumptions, and we should constantly ask ourselves whether they are reasonable, doubtful, or even utterly wrong. As an illustration of the caution needed before plunging into a complex calculation, consider another classic coincidence story that appears regularly on no-news days, which concerns someone who has bought a box of large eggs and found they all have double yolks. As usual, this leads to the question 'What's the chance of that?' In an example from 2010,[26] someone from the 'Egg Council' said that only 1 in 1,000 eggs were double-yolked, and so the chance of getting 6 such eggs in a box was claimed to be 1 in 1,000,000,000,000,000,000,000 ($\frac{1}{1,000}$ multiplied together 6 times).

The first step is to check whether this number is remotely plausible. There are around 2 billion (2,000,000,000) half-dozens of eggs sold in the UK each year, but even with this huge number we would only expect such a rare event to occur once in every 500,000,000 years. And it's just happened, so this immediately suggests the claimed probability is hopelessly wrong. An obvious error is the assumption that eggs in a box are independent, whereas they tend to come from the same flock and so getting one double-yolked egg increases the odds of getting another in the same box.

But maybe there is a more fundamental problem with our assumptions. I demonstrated this by going out and buying a box of eggs, cracking them open and finding they were all double-yolked! Extraordinary!

Or maybe not – Figure 4.3 reveals that I had bought a box labelled 'double-yolked'. It turns out that such eggs can be easily detected by holding them to a light and selected for inclusion in a box, and so the six eggs in the original story had presumably been sorted out and then put into an ordinary box, maybe after the order for double-yolked eggs had been fulfilled. We have no idea how often this happens, but it is certainly not once in 500,000,000 years.

This ludicrously trivial example illustrates an important point. We need to

Figure 4.3
I bought a box of eggs and they all had double yolks! But this was hardly surprising.

constantly question our assumptions and have the humility to acknowledge that the whole basis of our thinking may be incorrect.

———————

In this chapter we have looked at coincidences and shown that apparently surprising events are expected to happen surprisingly often. Our examples have sometimes been fun rather than important. But many rare and unexpected events will not make us smile, whether they concern financial crashes, environmental disasters, asteroid impacts, and so on, right through the long list of major harms we could face. Such catastrophes may be different from anything that has previously occurred, and so we require imaginative approaches to handling our uncertainty, as we shall see in Chapter 12.

Summary

- People are fascinated by coincidences – surprising concurrences of events that stay in our memory.
- Sometimes we can assess probabilities for coincidences to occur, particularly when they involve matches.
- It is important to distinguish between the small probability of a specific event to occur and the much larger probability of some similar event to happen at some point over a defined period.
- The Poisson approximation is useful, requiring only the expected number of events within a particular time period.
- Careful analysis is necessary to judge whether an apparently very unusual series of tragic events, such as deaths of medical patients, is proof of malicious behaviour.
- All analyses of surprising events are based on strong assumptions, and we should be vigilant about their plausibility.

CHAPTER 5

Luck

'Whether an exposed subject does or does not develop a cancer is largely a matter of luck; bad luck if the several necessary changes all occur in the same stem cell when there are several thousand such cells at risk, good luck if they don't. Personally I find that makes good sense, but many people apparently do not.'[1]

—Richard Doll, epidemiologist who helped
confirm the link between smoking and cancer

At around midday on 19 August 1949, under a cloud of thick mist, a British European Airways DC-3 (Dakota) on a flight from Belfast to Manchester flew into a hillside on Saddleworth Moor near Oldham in Lancashire.[2] All the crew and twenty-one of the twenty-nine passengers died on impact or soon afterwards. Eight passengers survived, including a young boy and his parents, although devastatingly their younger child was one of the fatalities. That surviving boy became my friend and statistical colleague, Professor Stephen Evans.

On Christmas Day 1971, 17-year-old Juliane Koepcke was on LANSA Flight 508 over the Amazon jungle when the plane was struck by lightning. She was thrown out, still strapped into her seat, and fell 3,000 metres. However, the thick jungle canopy broke her fall, and she survived, although ninety other people, including her mother, died.[3]

Our immediate response is to say that Stephen Evans and Juliane Koepcke were both very lucky. But what is good – or bad – luck? How much of your life's outcomes come down to luck? And how does *acknowledgement* of luck relate to *belief* in luck, perhaps better known as superstition? Although we routinely talk about luck, we never seem to ask ourselves what it actually means, or what role it plays in our lives.

What is 'luck', and what is its impact?

When we review events that have happened, we might say that someone has been lucky, or unlucky, if they have benefited or been harmed by something that is beyond their control, often perceived as being an unlikely chance event. Author and gambler David Flusfeder refers to luck as 'the operations of chance, taken personally'.[4]

The crucial element is the lack of control, although often when we look at 'lucky' outcomes, we realize that people may have had more control than is immediately obvious. For example, in a radio interview I did with Stephen Evans,[5] he told me that his father's experiences in the RAF led him to insist the family sat at the back of the plane – and the only survivors had been seated at the back. And Juliane Koepcke not only survived her fall, but also lived for eleven days in the jungle on her own, finding her way to an encampment and rescue – she only managed this because she had been brought up in the Amazon and had the necessary skills both to navigate and to look after her wounds.

There can also be major, but unpredictable, consequences of unlikely chance events that we might label as 'lucky'. Stephen Evans revisited the crash site and reported having an 'immense sense of gratitude for being alive and reflected on many good things that have come out of what was obviously a tragedy for our family and much worse for many others'. After a lengthy recovery from his injuries, the compensation enabled him to have a strong education and lead what he felt was an enormously privileged life.*

Perhaps the most notorious example of the impact of a chance event involves Gavrilo Princip, who was one of a team of assassins waiting for Archduke Franz

---

* Stephen describes himself as a follower of Christ, which will of course affect his view of his good fortune. He also recounted how he was once sitting next to a very nervous passenger during an aborted plane landing and tried to be reassuring: 'I said to her, what are the chances of someone being in two plane crashes, and she said millions to one against. I said, well, don't worry, I have been in one. But it did not have the desired effect – she was quite convinced that I was the cause of the first and about to be the cause of this next one.' Perhaps she understood conditional probability.

Ferdinand in Sarajevo in June 1914. After a failed initial attempt they abandoned their mission, but later that day the archduke's driver took a wrong turning and stalled the car right in front of Schiller's delicatessen, where Princip happened to be standing. Princip reacted rapidly and killed both the archduke and his wife, which was perhaps lucky for Princip, but unlucky for his victims, and for the millions who were dragged into the subsequent world war.

In contrast, Winston Churchill had a lucky escape in December 1931 when he was knocked down and seriously injured in New York after looking the wrong way when crossing Fifth Avenue.[6] In an almost contemporary but sadly unverified parallel event, John Scott-Ellis (later Lord Howard de Walden) told the story of driving through Munich in August 1931, a few months before Churchill's accident, and knocking someone down – he later claimed this was Adolf Hitler.[7] Needless to say, the course of history might have been rather different if either of these near misses had been fatal.

Are there different types of luck?

Since Aristotle, philosophers have argued about whether people should be praised or blamed for events that occur outside their personal control – so-called moral luck.[8] A classic thought experiment concerns two friends Alan and Bill who go to the same party, get equally drunk, drive the same route home, but Bill is suddenly confronted with a child who steps in front of his car and is killed. Which of the two friends deserves more blame? Bill would tend to be judged more harshly than Alan, but in a sense they were equally culpable, and Alan was just lucky.

These arguments are beyond the scope of this book, but have given rise to a useful categorization of types of luck.[9]

- *Resultant luck*, where individuals are in similar situations but some have good and some have bad outcomes due to factors beyond their control. Such as winning the lottery, surviving a First World War infantry attack, or Bill and Alan driving home.
- *Circumstantial luck*, in which the vital factor deciding the outcome is being in the right place at the right time, or in the wrong place at the wrong time – such as in a plane about to crash.
- *Constitutive luck*, which is a property of the person you were born as – your parents, background, genes and character traits.

Stephen Evans experienced all of these. He had good resultant luck in surviving the crash, bad circumstantial luck in being on the plane, and good constitutive luck in having sensible and caring parents.

Others are not so fortunate. Consider what happened to 55-year-old Felicity Chilcott on 9 January 1951. She was on a bus to her home in Cumberland Mews near Regent's Park, unaware that a chimpanzee called Mr Cholmondeley (pronounced 'Chumley') had meanwhile escaped from the sanatorium at Regent's Park Zoo. Buses then had open rear access, which allowed Mr Cholmondeley to board the bus. He then bit Felicity twice on her leg.

This story* combines the bad luck of being on the wrong bus at the wrong time, and being the one that Mr Cholmondeley happened to pick on. Maybe remember this when you feel you've had a bad day.

We can go even further back in considering our fortune. *Existential luck* arises through just being born; Buddhism teaches that being reborn as a human is as rare as a turtle surfacing once every hundred years and happening to put its head through a single golden ring. Then we might consider the luck that life arose on our planet, or that our solar system exists at all. Although, given that we do exist, it is perhaps questionable whether all these existential concerns make sense (see Chapter 16).

Luck is not necessarily to do with good or bad events happening to us. We can have *epistemic luck* if we have a correct belief but for the wrong reasons; for example, if we claim ten out of ten confidence in an answer to the quiz in Chapter 2, and this turns out to be correct, but we only had this belief because we misunderstood the question.

To my mind, constitutive luck is the most important. You have no control over the situation of your birth and your early upbringing, and yet it is clear that these factors have an overwhelming influence over the trajectory of your life. I have personally benefited from extraordinary constitutive luck, with my kind parents, robust health, and being born into a period and context which was peaceful, full of opportunities and with good post-war government support.

When attributing reasons for any success they have, people tend to

---

* The chimpanzee had been in the zoo since 1948, performing at the then-popular 'Chimps' Tea Parties', and he was being treated for a cold. In 1951 the 3 and the 53 bus went up Albany Street towards Cumberland Mews.

overestimate the role of their efforts and acquired skills, whereas they should mainly be grateful for being dealt a good hand at birth – their constitutive luck.

————————

Sometimes our luck is not obvious at the time. My grandfather Cecil Spiegelhalter had the bad constitutive luck of being a man born in the 1880s, just ready for the First World War, and even worse circumstantial luck of being gas officer in the Ypres sector. He probably did not think himself fortunate when a German shell landed near him on 29 January 1918, as I recounted in the Introduction, but it turned out that this resultant luck prevented his being sent to one of the most dangerous places in the First World War.

Or consider the astronauts on the early missions of the US Space Shuttle. On the fleet's twenty-fifth mission on 28 January 1986, the Space Shuttle *Challenger* exploded in front of millions of viewers, killing all seven crew. Richard Feynman was a member of the Rogers Commission on the *Challenger* accident, and after graphically illustrating the effect of sub-zero launch temperatures on the flexibility of the O-ring seals,* wrote in a personal appendix to the commission's report: 'It appears that there are enormous differences of opinion as to the probability of a failure with loss of vehicle and of human life. The estimates range from roughly 1 in 100 to 1 in 100,000. The higher figures come from the working engineers, and the very low figures from management.'[10] But even Feynman may have underestimated the risks that the crews were facing.

In 2011, the NASA Space Shuttle Safety and Mission Assurance Office carried out a retrospective risk analysis of 135 Shuttle missions over thirty years, explaining in a radio interview that 'we're taking our current knowledge that we have today, and applying that to the configuration of the vehicle back then'.[11] They concluded that the risks had been substantially greater than assessed at the time, with the first launches having a roughly 10% probability of catastrophic loss, ten times as high as Feynman's worst figure.[12] They estimated only a 6% probability of reaching the twenty-fifth launch (*Challenger*) without losing a craft, and concluded: 'We were lucky, there were a number of close calls.' The early Shuttle crews did not realize how lucky they had been.

————————

* At a televised hearing on 11 February 1986, Feynman put a piece of O-ring material in a glass of iced water and showed how this immersion made it lose its flexibility.

The order of play of international cricket is decided by a coin toss, and when he was captain of the England cricket team, Nasser Hussein managed to lose fourteen tosses in a row.[13] This was considered bad luck, with a probability of $\frac{1}{2}^{14}$ = 1 in 16,000, but what about this apparently remarkable event?

> In a TV programme, illusionist Derren Brown was filmed flipping ten heads in a row. Was he lucky or unlucky?

This occurred in *The System*,[14] and Brown admitted later in the programme that the segment shown was taken at the end of *nine hours* of filming him trying to complete this task – a fine example of deceiving the viewer by carefully selecting what is shown. But was he lucky or unlucky to have taken so long to get his ten heads?

This requires analysing the time it takes for an event to occur. Consider the simple task of throwing a die until you get a six. How many throws will you need? We can assume each throw is independent, and the die is nicely symmetric, and so if I have thrown the die without success so far, the probability of getting a six on the next throw is $\frac{1}{6}$. So what is the probability that I will get my first six on, say, the third throw? This means I did not get a six on the first two throws (which has probability $\frac{5}{6} \times \frac{5}{6}$) but then I do get a six on the third throw (with probability $\frac{1}{6}$) so the total probability is $\frac{5}{6} \times \frac{5}{6} \times \frac{1}{6} = \frac{25}{216} = 0.12$. This is shown in Figure 5.1, together with the probabilities of all the other possibilities.

This is known as a **geometric distribution**.* The mean average of the distribution in Figure 5.1 is 6, so that if I want to throw a six, on average I will take 6 throws, and if I want to throw 100 sixes, I can expect to take 600 throws. The useful general rule is that if we are trying to achieve something, with a chance p of succeeding at each independent attempt, then on average we will need $\frac{1}{p}$ attempts to achieve it, although there is 37% chance that we will need more.† The **median** of the distribution is 4, since in 51% of attempts,

---

* If each independent event has probability p, then the probability that the first event occurs at attempt x is $(1 - p)^{x-1} p$. This geometric distribution has expectation (mean average) $\frac{1}{p}$. For small p, the distribution is approximated by an **exponential distribution** with mean $\frac{1}{p}$, so that $\Pr(X > x) \approx e^{-xp}$.

† The probability of taking more than the mean number of attempts is $\Pr(X > 1/p) \approx e^{-p/p} = e^{-1} = 0.37$.

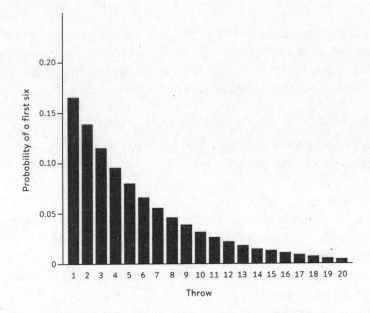

Figure 5.1
The geometric probability distribution for the first die
throw on which a six occurs. The distribution has mean 6,
median 4 and mode 1.

I will need 4 or fewer throws to get a six – this perhaps can be thought of as
the 'typical' number of throws needed.

The **mode** of the distribution – the most likely throw to get the first six – is
the very first throw. This can seem rather unintuitive, but with a constant chance
of occurring at each attempt, the most likely time for the next event is straight
away. This can help explain why apparently random events tend to cluster, as we
saw in Chapter 4. If unconnected rare occurrences, such as radioactive decay or
(most) plane crashes, have a constant chance of happening in each unit of time,
then the gaps between them will have an exponential distribution. There is no
force that will encourage them to be equally spaced – after a plane crash, the
most likely time for the next one is immediately.

Now back to Derren Brown flipping his coin for nine hours. If we consider
an 'attempt' as flipping a coin until we get a tail, then the probability of an
attempt successfully getting 10 heads in a row is $\frac{1}{2} \times \frac{1}{2} \dots \times \frac{1}{2}$ (10 times)
which is $\frac{1}{1,024}$. Therefore, from what we know of the geometric distribution,
an average of 1,024 attempts will be needed.

We can estimate that he made around 1,600 attempts before getting to 10 heads,* and so Brown appears to have taken longer than average. There is around a 21% chance that it would take as long as this before he succeeded,† so he was a bit unlucky, but not badly so. And it was extremely impressive that even after nine hours of filming he could appear as if it were his first attempt, and keep cool as he got closer to ten heads, which must have been a great relief.

My colleague James Grime re-enacted this feat on film[15] and took only an hour, succeeding on attempt 234. The chance of achieving this so quickly is around 20%, so he was as lucky as Derren Brown was unlucky. James valiantly decided to carry on to 1,024 attempts, which took him another five hours of flipping, and he didn't manage to repeat his success. This is serious dedication to demonstrating the play of chance.

———

Winning numbers in lotteries are chosen at random, and so when ball 39 made the most appearances in 2022,[16] it would have been reasonable to label 39 as the 'luckiest lotto number' (although we shall see in Chapter 6 that 39 was the *least* lucky number in the first fifty lottery draws). But unlike the lottery, sport is not supposed to be just a matter of chance, and so it may seem strange that researchers have asked questions such as

> How much of the variation in top football leagues comes down to luck?

In a standard season for a football league, each team will play each other team twice – once on the home pitch and once away. A win will score a team three points, and a draw gives each team one point each. As the season progresses the points are added up, and the English Premier League 2022–3 season produced the final distribution of points shown as grey blocks in Figure 5.2. Superimposed is a curve showing what we would have expected the distribution to be, if matches were luck alone.

---

* In the filmed sequence he had around a seven-second introduction and each flip took on average seven seconds. Each attempt ends on the first tail, which again from the geometric distribution means that each attempt lasts on average two flips, which would imply an average of around twenty seconds for each attempt. Nine hours is 540 minutes, and so he took around $540 \times 3 \approx 1,600$ attempts.

† The probability of taking at least 1,600 attempts is around $e^{-1600/1024} = 0.21$.

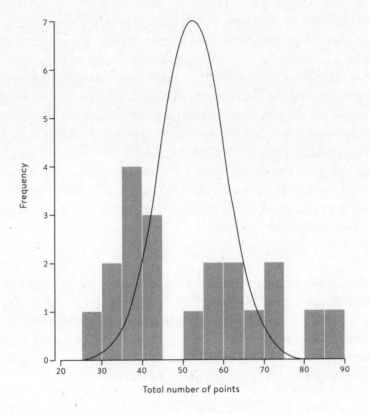

Figure 5.2
Grey blocks represent the distribution of points scored in the
2022–3 English Premier League; Manchester City were top with
89 points, while Southampton were bottom with 25. The smooth
curve shows the distribution that would be expected if the results
of each match were decided by chance alone, according to the
observed proportions of home wins, draws and away wins.

So how do we work out what we would expect this distribution to look like
if the results were decided by pure chance?[17] In top football leagues, roughly
50% of matches are won by the team playing on their home ground, around 25%
are a draw, and about 25% are won by the team playing away – we can call this a
50/25/25 breakdown. Suppose at the start of each match, instead of a coin flip
just deciding the direction of play, it actually decides the match. If it comes up
heads, the home team wins. If it comes up tails, it is flipped again, and this time
if it comes up heads the match is declared a draw, and if tails the away team wins.

This would save a lot of time and effort, and although it may not make riveting watching for a crowd or a television audience, it would end up with the correct 50/25/25 breakdown that is observed in practice. All the teams would be essentially equal since the games are decided by chance, but at the end of the season there would still be a full league table of points, with a team at the top and one at the bottom – someone has to be top, even though it's just luck who it happens to be. An alternative thought experiment that should ensure equality of the teams would be for each team to be chosen each week at random from the pool of all players in the Premier League.

Were the games decided at random, using the actual proportions observed in the Premier League in 2022–3 which was 48% home wins, 23% draws and 29% away wins, we can estimate what we would expect the distribution of points to be, which is shown as the smooth curve in Figure 5.2.

At the end of the season some teams are clearly outside the 'chance' distribution, and so we can conclude there are genuine differences between the teams, although a substantial proportion of the spread of the final points is explainable by chance alone. There are different ways of summarizing this proportion; for example, we can say that 45% of the standard deviation of the Premier League points is due to chance or luck. More simply, we can look at the observed spread of points (25 to 89, or a range of 64) and compare it to the spread we would expect from matches decided by chance (39 to 66, or a spread of 27). So 27/64 = 42% of the observed spread is explained by chance, which is nearly half.

Table 5.1 shows this analysis repeated using the 2022–3 results for the major European leagues,[18] ordered by the final column – the proportion of the observed spread that is explained by chance or luck.

Those teams at the top of the table, with the lowest proportion of spread explainable by chance, tend to have a few outstanding teams that dominate the league, such as Rangers and Celtic in the Scottish Premier League. In contrast, those leagues at the bottom of the table are much more evenly matched – nearly two thirds of the final point-spread in the Scottish Championship (their second league) is explainable by chance.

It is notable how the lower half of the table, where teams are more equally matched, is almost entirely composed of second-division leagues. The exception is the German Bundesliga 1, which shows 58% of its point spread is explained by chance, indicating remarkable similarity between the top-flight German teams.

The English Premier League is about midway down the table, which is perhaps the 'Goldilocks' position, where there is sufficient difference between the teams to produce some clear winners, while not different enough to make

League	No. teams	% home wins / draws / away wins	Bottom team (points)	Top team (points)	Observed spread	Expected spread if just chance/luck	Proportion observed spread explained by chance/luck
Scotland: Premier League	12	52 / 18 / 31	Dundee United (31)	Celtic (99)	68	42 to 65 = 24	$\frac{24}{68}$ = 35%
Greece: Ethniki Katigoria	14	42 / 25 / 32	Apollon (21)	Olympiakos (83)	62	36 to 39 = 23	37%
Italy: Serie A	20	42 / 26 / 31	Sampdoria (19)	Napoli (90)	71	39 to 65 = 27	38%
Netherlands: Eredivisie	18	45 / 24 / 31	Groningen (18)	Feyenoord (82)	64	35 to 59 = 25	39%
Portugal: Liga 1	18	49 / 18 / 33	Santa Clara (22)	Benfica (87)	65	35 to 61 = 26	39%
France: Le Championnat	20	43 / 24 / 33	Angers (18)	Paris St Germain (85)	67	39 to 66 = 27	40%
Turkey: Ligi 1	19	45 / 21 / 34	Hatayspor (23)	Galatasaray (88)	65	37 to 63 = 26	41%
Spain: Liga Primera	20	48 / 23 / 29	Elche (25)	Barcelona (88)	63	39 to 66 = 27	42%
England: Premier League	20	48 / 23 / 29	Southampton (25)	Manchester City (89)	64	39 to 66 = 27	42%
Belgium: Jupiler	18	44 / 22 / 34	Seraing (20)	Genk (75)	55	35 to 60 = 25	46%

League	Teams	% H/D/A	Bottom team (points)	Top team (points)	Observed spread	Expected spread	Ratio
England: Championship	24	41 / 27 / 32	Blackpool (44)	Burnley (101)	57	47 to 78 = 31	54%
Italy: Serie B	20	41 / 32 / 27	Benevento (35)	Frosinone (80)	45	38 to 64 = 26	57%
France: Division 2	20	44 / 27 / 29	Niort (29)	Le Havre (75)	46	39 to 65 = 26	57%
Germany: Bundesliga 1	18	47 / 25 / 28	Hertha (29)	Bayern Munich (71)	42	35 to 59 = 24	58%
Spain: Liga Segunda	22	44 / 33 / 23	Lugo (31)	Granada (75)	44	43 to 69 = 27	61%
Germany: Bundesliga 2	18	46 / 24 / 30	Sandhausen (28)	Darmstadt (67)	39	35 to 59 = 25	63%
Scotland: Championship	10	39 / 29 / 31	Cove Rangers (31)	Dundee (63)	32	38 to 59 = 20	64%

Table 5.1

Analysis of major European football leagues for the 2022–3 season, showing the % of home wins, draws and away wins, the bottom and top teams and their points, the observed spread in points, the expected spread were the games decided by chance, and the ratio of the expected to the observed spread. Teams are ordered by the final column – the proportion of the spread of points explained by chance.

the games too predictable. But the Premier League used to be much closer fought before the massive injection of funds for some teams: for example in the 1996–7 season, Manchester United was top of the table but had only 75 points, and Nottingham Forest was bottom with 34, with 62% of this spread explainable by chance, similar to current second divisions.

How can we assess the 'quality' of each team?

The end-of-season point total describes how well each team has performed; for example, in the English Premier League 2022–3 season, Manchester City played 38 games and were awarded a total of 89 points, an average of $89/38 = 2.34$ points per game, compared to Southampton's $25/38 = 0.66$. This could be viewed as an estimate of some abstract measure of underlying 'quality' – with some imagination, this can be thought of as the average points each team would score per game were the season to continue indefinitely; this concept of an underlying 'true' average points per game is analogous to the idea of a player's 'true skill' estimated from video-game competitions.[19] We can calculate a margin of error for this estimate, and so put an uncertainty interval around each observed average points per game, as shown in Figure 5.3.

The substantial overlap between the teams means that insufficient matches have been played to be confident about the 'true' quality of each team. We can, however, be very confident about five teams being greater than average (the dotted line), and five teams being worse than average.

Using the same ideas, we can even explore our uncertainty about the 'true' rank of each team, which can be thought of as where they would end up in an imaginary league table were the season to continue indefinitely.

Figure 5.4 shows there is huge uncertainty as to the true ranks of the teams. There are five teams we can be reasonably sure are in the upper half, while only four teams can be confidently placed in the lower half, including the three bottom teams that were relegated.

We can also investigate the probability that the season's winner, Manchester City, really was the best team. We assess this to be 67%, compared to Arsenal's 27%, which could be interpreted as the probability that Arsenal would actually end up top of the league table, were the season to continue indefinitely. And were the teams that were relegated really the three worst teams? The probability of truly being in the bottom three for 'quality' is judged to be 88% for Southampton, 58% for Leeds, and 47% for Leicester.

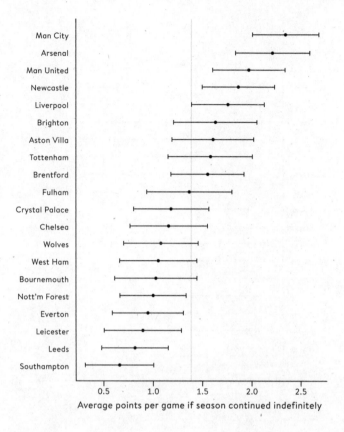

Figure 5.3
The dots are the observed average points per game for each English Premier League team at the end of the 2022–3 season. The ranges show the 95% probability interval for where the eventual 'average points per game' would lie if the season continued indefinitely. The grey line down the middle is the average of the averages.

Everton narrowly escaped relegation, and in fact we assess they had a 28% probability of truly being in the bottom three teams.

These techniques for putting uncertainty around the ranks in a league table have been applied to IVF clinics,[20] surgical units and schools.[21] Such analyses give a healthy and sceptical perspective on league tables; there is generally huge uncertainty about rankings, and just because someone comes top, it does not necessarily mean they are definitely the best – we know that a substantial amount of luck has been involved in their success.

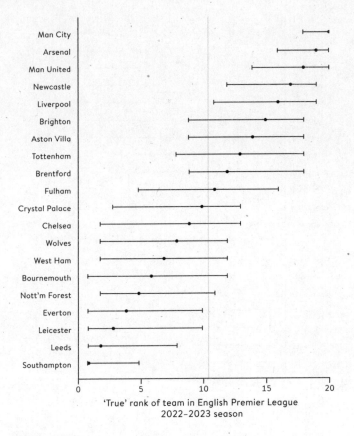

Figure 5.4
The dots are the observed ranks for the teams at the end of the season. The ranges show a 95% probability interval for their eventual rank were the season to continue indefinitely.

Luck in the future

So far we have looked at past events and judged how much luck was involved in the way things turned out. But many people appear to believe in luck as an external force influencing the future. This means we are now venturing into the murky territory of superstition, auspicious occasions, omens, fetishes, amulets, crystals, astrology, numerology, and so on. I don't intend to take up precious space on these topics.

Of course, some actions may appear to be mere superstition, but may be playing a useful practical role. For example, many sporting stars routinely perform rituals. Before every ball received when batting, test cricketer Ed Smith wiped his forehead with his thumb, touched the peak of his helmet, pushed down the Velcro of each glove and rearranged his thigh pad. He might do this up to 200 times in one day, and has said, 'I suppose I was doing something with all that scratching and fiddling around before each ball.' Presumably he was keeping his focus during the repeated gaps in intense activity, although he is ready to acknowledge the role of luck in his life – he met his wife on a train.[22]

Even if we do not believe in some mystical process that will bring us good fortune, can we improve our chances of events turning in our favour? The first step must be to improve our knowledge and skills, following the dictum (generally attributed to golfer Gary Player) that 'The more you practise, the luckier you get.' Apart from this obvious tactic, psychologist Richard Wiseman has concluded that so-called lucky people tend to follow four basic principles.[23]

- They spot and take advantage of any opportunities they encounter.
- They have good intuition about appropriate actions.
- They have positive expectations, giving them confidence to act.
- They have the resilience to deal with adverse events and turn them to their advantage.

Juliane Koepcke amply demonstrated these traits when she survived her jungle journey. Similarly, people who make unplanned discoveries – by the process called 'serendipity' – have not just been lucky. In 1928 Alexander Fleming discovered penicillin after spotting that some mould seemed to be preventing the spread of bacteria in a dish that had been left during a holiday, but he was engaged in an intense research programme on combating bacteria. It then took years of effort to develop useful antibiotics, with the pharmaceutical company Pfizer enabling mass production of penicillin during the Second World War.

Pfizer later experienced its own serendipitous discovery when investigating treatments to dilate heart blood vessels, when Welsh miners recruited to a clinical trial reported that sildenafil had an unexpected effect on the blood flow through another part of their anatomy. The company realized it could rebrand the compound as a treatment for erectile dysfunction, and Viagra was born. Sildenafil was later also approved for treating pulmonary arterial hypertension, vindicating the original plan.

In each of these cases, the discovery was part of a prolonged investigation, and recognizing and exploiting the surprising observations required imagination, insight and investment. So not just luck.

How can we sensitively communicate the role of 'luck'?

A few years ago I was involved in building a website about children receiving heart surgery[24] which was being developed in close collaboration with affected families. The statistics about survival were explained by saying, for example, 'of 100 children having this operation, we would expect 98 to survive and, sadly, 2 not to survive', but we struggled to find words for the reasons why some children survive and some do not. Could we call it luck? Chance, fortune, fate? These all seemed very insensitive. Technical terms like 'binomial variation' and 'random variability' were even worse. 'Unavoidable unpredictability' came closest, but it felt too clumsy.

We canvassed opinion on this, and in the end a student suggested we could say that some babies would die because of 'unforeseeable factors'. We immediately accepted this term, and it was well accepted by parents. So that's the phrase I now use in all such circumstances.

––––––––

After considering all the stories in this chapter, my personal feeling is that luck is not some mystical force – that your real luck happens when you are born, and after that it's a matter of making the best of the hand you've been dealt in the face of uncontrollable outside events. So if we think of chance as unavoidable unpredictability, then I agree that 'luck is chance, taken personally'. Events get labelled as 'lucky' in hindsight; Ed Smith happened to start a conversation with a woman on a train, but presumably this only became labelled as luck in retrospect, particularly when they decided to get married.

I started this chapter with a quote from Richard Doll about the role of luck in getting cancer. This seems a classic case of bad *resultant*, or outcome, luck – a stem cell just happens to develop a particular mutation and sets off the first stage of a tumour – but some people have the bad *constitutive* luck of being born with a genetic defect that causes cancer. This was brought home to me when it happened to my son, and may be a reason why I don't search for explanations for everything that happens and try to accept luck and uncertainty as it plays out, for good or ill. As the story at the start of the next chapter will illustrate.

Summary

- We can label the operations of chance as 'luck', and such chance events can have major consequences.
- Philosophers have identified 'resultant', 'circumstantial' and 'constitutive' luck. Perhaps the most important form of luck concerns the circumstances of your birth, over which you have no control.
- Sometimes we can quantify the amount of luck in past events.
- By examining the role of chance in league tables, we can assess the uncertainty about the 'true' rank of each team or organization.
- Many people appear to believe in luck as an active force influencing future events. Even without this belief, some behaviours and attitudes are linked to people being perceived as 'lucky'.
- Care is needed in communicating sensitively the role of chance in important events.

CHAPTER 6

It's All a Bit Random

In 2016, I was diagnosed with prostate cancer, adding to an extensive history of cancer among my close family members. As a father of two daughters, I was concerned that I might be a carrier of the BRCA2 gene mutation that markedly increases the chance of developing breast and other cancers. I presented my family's experiences to a genetic counsellor, who entered it into the specialized software called BOADICEA that assesses probabilities of having specific genetic abnormalities based on family histories of cancer.[1] It came out with an estimated probability of 33% that I was carrying the damaged BRCA2 gene, which was more than 100 times the background rate, and sufficiently high to get me tested.

I either have the gene or I don't, and so this '33% probability' is a classic example of epistemic uncertainty, just like a flipped coin that has been covered up. But underlying this number lie further probabilities, including assumptions about the distribution of the gene in the population and its relationship to the risk of different cancers. Crucially, the probability depends on assumptions about *Mendelian inheritance*, which prescribes your fixed chances of inheriting a gene from your parents.

Every human has twenty-three pairs of chromosomes, each a long DNA molecule, in the nucleus of a cell. Genes are a site on a pair of chromosomes, where individuals have a pair of versions of the gene known as alleles. In normal human reproduction, the offspring arises from an egg and a sperm, each of which contains one of the parent's pair of chromosomes (or a combination of the two).* So at a particular site, there is a 50:50 chance of an offspring inheriting either allele from each parent, like flipping a coin.

Figure 6.1 shows a basic model for the inheritance of cystic fibrosis (CF), a disease which only occurs when both versions of the CFTR genes are 'CF', although around 1 in 25 of the population have just one CF gene and are known as carriers. The basic symmetry of inheritance means that the probabilities of the different possible outcomes can be easily calculated.

Similarly, if one of my parents carried the BRCA2 gene mutation on either chromosome, there is a 50% chance that I would have inherited it, and this assumption contributes to the calculated 33% probability of being a carrier of the mutation.

I was relieved, for the sake of my daughters, to find I was not a carrier of the BRCA2 mutation. But for me it raised the question – does this '50% chance' of inheritance really mean there is some truly random mechanism that decides which chromosome I inherit? Or is it just a very complex but mechanical process whose symmetries mean the chances balance out at 50:50?

———

So much of what happens in the natural world is uncertain. The way molecules of water create currents and waves in the sea, the ice crystals that form an uncountable variety of snowflakes, and the precise characteristics of everything that lives: all are unpredictable. But how much of this is true randomness, sometimes known as **stochastic**? And how much is due to the fact that a system is so complex that huge numbers of small influences tend to produce variation indistinguishable from randomness, even though the process is entirely deterministic? Or maybe it is a so-called **chaotic system** in which tiny differences in initial conditions are amplified and mimic randomness, well known from the classic image of a butterfly flapping its wings possibly causing a tornado weeks later. Is the unpredictability due to 'chance', or due to 'complexity'?

---

* Each of the mother's cells has 23 pairs of chromosomes, and each of her eggs has a copy of one of each pair, apparently giving $2^{23} \approx 8,000,000$ different possible combinations. But reality is even more complex, since around half of the chromosomes in the egg will not be an exact copy, but a combination of the pair which have split and rejoined at some site in the chromosome. And the same holds for the father's sperm.

The goddess Fortuna, or whatever you might believe in, happened to choose you from an essentially infinite number of possibilities. No wonder siblings can turn out so differently.

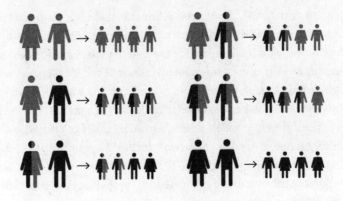

Has CF   Carrier   Not a
                     carrier

Figure 6.1
A person who has only one CF gene is called a CF carrier, and does not
have the disease but can pass it on. If both parents have the disease,
then their child is certain to inherit two CF genes and so have CF. If both
parents are carriers, there is a 25% chance that their child will inherit
their CF genes and have the disease, a 50% chance that the child will
become a carrier, and a 25% chance of not inheriting either CF gene and
so not even being a carrier. If one parent has the disease, and one is a
carrier, there is a 50% chance of having CF and a 50% chance of being a
carrier. The figure shows what we would expect if each combination of
parents had four children. Adapted from the Cystic Fibrosis Foundation.[2]

This is a big question. For example, evolution occurs when the genome in the
cells of a parent mutate and those mutations are passed to offspring. If the environment for some reason favours individuals with those mutations, they tend
to produce more offspring, and hence the mutation is passed on and amplified
in successive generations.* The mutations occur because of microscopic influ-

---

* Each base-pair in the genome has around a 1 in 500 million probability of mutating each year, and there are
about 3 billion base-pairs, so we might expect around 6 mutations per year.

ences from external factors or errors in cell replication, and so are impossible to trace back to specific hidden causes. So evolution cannot be clearly labelled as either chance or complexity, stochastic or deterministic.

Similarly, epidemiologists such as George Davey-Smith have identified that most of the variability between people's health over their lives is unexplained by any measurable risk factors, including their genes. Genetically identical individuals can end up with dramatically different and unpredictable health outcomes – I previously quoted Richard Doll as saying that whether someone got cancer or not is largely luck. Writer and broadcaster (and founder of *More or Less*) Michael Blastland refers to this as the *hidden half*, as in the phrase 'you don't know the half of it', representing 'enigmatic variation – the many mysteries and surprises that humble human understanding'.[3]

So what underlies this extraordinary unexplained variability? We are building up to the big philosophical question . . .

Is the world fundamentally deterministic or stochastic?

When writing about probability in 1814,[4] French genius Pierre-Simon Laplace imagined 'an intellect which at a certain moment would know all forces that set nature in motion, and all positions of all items of which nature is composed, if this intellect were also vast enough to submit these data to analysis, it would embrace in a single formula the movements of the greatest bodies of the universe and those of the tiniest atom; for such an intellect nothing would be uncertain and the future just like the past could be present before its eyes.'

In other words, if we were some omniscient being that knew everything about the current world, and all the laws that governed it, then the future could be predicted exactly. This thought experiment has become known as *Laplace's demon*, and represents extreme *determinism*, in the sense of believing that things happen according to fixed mechanistic laws; Laplace argued that the idea of probability was needed to deal with our personal ignorance of the massive complexity of a clockwork universe.

Then around a hundred years ago, quantum mechanics came along and apparently destroyed this argument. The work of physicists Niels Bohr, Werner Heisenberg and others concluded that, at the deepest subatomic level, the world is fundamentally stochastic – particles only have a probability distribution of possible positions and velocities, until they are observed, and then this distribution collapses to a single point.

One consequence of this fundamental uncertainty is the unpredictability of radioactive decay, when the nucleus of a large and unstable atom spontaneously disintegrates for no apparent reason, emitting a particle and leaving a (generally more stable) reduced atomic structure. The probability of a particular atom decaying in a fixed period of time is in general unaffected by the age of the atom, the temperature, or any external phenomenon, and so could be considered as a *determined probability*, an objective property of the world that is unconditional on the observer or anything else, so you could set your watch by it – which of course we do when using 'atomic clocks' based on resonant frequencies of caesium atoms. This is in distinct contrast to the subjective probabilities otherwise emphasized in this book.

There have always been arguments against this idea of irreducible, determined probabilities. Referring to God, Einstein commented: 'I am convinced that He is not playing dice.' There have been continued theories of 'hidden variables' that lurk behind particles and control their future state, although a Nobel Prize has been awarded to a team that claims to have disproved this idea.[5] There is also the 'many-worlds interpretation', or multiverse, in which everything that can happen does happen, and we just find ourselves in one of the outcomes.

But if we accept the mainstream opinion about the quantum world, it is natural to ask whether this essential stochasticity at the heart of matter influences what we actually observe in our lives. The general view is that 'quantum indeterminacy' will average out when it comes to things as big as molecules, and certainly for biological cells, but others claim that even coin flips might be influenced by quantum effects in the brain.[6] This would be remarkable, as it would connect subatomic randomness to chance as we actually experience it.

While all these debates are doubtless fascinating and important, they are well beyond my pay-grade, and you should read elsewhere if you are interested. Fortunately, since I define uncertainty as a relationship between an observer and an event, I can neatly sidestep having to have an opinion on whether the huge and unexplained variability between biological organisms (such as people) is due to genuine randomness, or deterministic but unknown influences. It simply makes no practical difference to me. Indeed, the world may be totally governed by the will of God, and so completely predestined, but as we don't know that will, we are still left with our epistemic uncertainty.

But even if we can avoid arguing about whether the world is *truly* stochastic or deterministic, we still need to decide how to treat different phenomena.* It is helpful to start at the most basic level and steadily increase in scale.

- Subatomic particles are generally assumed to be *stochastic*.
- Individual molecules can be assumed to be *deterministic*, obeying Newton's laws of mechanics.
- As soon as there are more than two molecules, we cannot calculate their relative movements, and outcomes are immensely sensitive to initial conditions. So the behaviour of lots of gas molecules can be treated as if they are *stochastic*, reflected in the theory of statistical mechanics.
- Larger bodies of gas or solids obey Boyle's and Newton's laws, and so can be treated as *deterministic*.
- The behaviour of individual organisms or people is considered as *stochastic*, as in genetics.
- Large groups of people become (almost) *deterministic* in some respects, such as the number of suicides being broadly predictable. This became known as 'statistical fatalism' in the 1800s, when researchers such as Adolphe Quetelet revealed apparent 'laws' that governed group behaviour, which is essentially the regularity of a Poisson distribution.
- The development of societies has an unpredictability that perhaps is best treated as *stochastic*.

So as we expand our vision from the subatomic to whole societies, it turns out that there is a repeated pattern of treat-as-stochastic microlevel events aggregating to give a regularity that can be treated deterministically, which

---

* Laplace's deterministic Demon would be countered by either true randomness, or the contentious topic of individual free will. Fortunately, we can avoid long-running arguments about whether free will is an illusion, since this is not dependent on whether the world is stochastic or deterministic. But, for what it's worth, my own view accords with Spinoza, who said that 'men believe themselves free, simply because they are conscious of their actions, and unconscious of the causes whereby those actions are determined', and this seems to be backed up by modern neuroscience arguing that we tend to make decisions *before* any conscious deliberation. So personally, I am happy to believe that my thoughts, intentions and actions arise through immeasurably complex chains of causes, rather than there being some autonomous free 'self'. But this is definitely getting off-topic.

then become treated as stochastic at larger granularity, and so on. At each level we are assuming a **model** for the world that helps to deal with the task in hand – such models are not reality but, as we shall see in Chapter 8, can be very useful.

My pragmatic approach means that when I talk about something being random, I mean that it is **effectively random**, in the sense of being practically indistinguishable from something that is drawn from some known probability distribution. And such effective randomness has a lot of applications.

Why did the Manhattan Project require random numbers?

Plutonium-239 is a man-made isotope with, when left to its own devices, a half-life of 24,100 years, which is the period over which we would expect half a set of atoms to decay to uranium-235, or equivalently the period over which there is 50% probability of a particular atom decaying. This sounds fairly stable but, if a critical mass is present, the particles emitted when one atom decays can trigger a decay in nearby atoms, leading to a chain of disintegrations and a huge release of energy. During the Second World War Manhattan Project – the atomic-bomb programme at Los Alamos in the US – scientists were struggling with mathematical solutions to modelling such chain reactions in a critical mass of radioactive material. It was just too complicated.

Stanisław Ulam, who along with Edward Teller would go on to design the first hydrogen bomb, was one of the scientists. A brilliant Polish-Jewish nuclear physicist, he was fortunate enough to have left Poland for the US on 29 August 1939, just three days before the Nazi invasion. Ulam loved games of chance, and enjoyed playing Solitaire (a form of Patience), and he tried to use his elegant mathematics to work out the probability of winning such games. He failed, but grudgingly realized that he could use brute force: play the game a hundred times with 'random' shuffles, and simply count the proportion of times that the game could be finished. He then made the brilliant leap to apply these 'statistical simulation' methods to understand complex atomic chain reactions, by repeatedly modelling single imaginary chain-reaction processes and looking at the proportion of times that a critical limit was reached. Apparently Ulam had an uncle who would borrow money so he could 'go to Monte Carlo' to gamble, and so the **Monte Carlo method** was born, in which complex calculations are replaced by repeatedly simulating possible sequences of events.*

The scientists at Los Alamos found that if a canister contained many small lumps of Plutonium-239, and these were then imploded suddenly into a single piece of around 6kg, there would be a chain reaction and the decay would no longer follow its standard slow pattern. Which is just what happened when Fat Man, the second nuclear bomb to be dropped on Japan, exploded over Nagasaki at 11.02 a.m. on 9 August 1945. Around 1kg of Plutonium-239 disintegrated in a fraction of a second, releasing energy equivalent to about 21,000 tons of TNT – approximately 35,000 people were killed in the immediate aftermath, and roughly the same number died from injuries and radiation exposure later.

Under normal conditions, it would have taken that 1kg (16% of the total) around 6,000 years to decay.† This sobering example demonstrates that although the probabilities of atomic decay are generally considered as 'objective', they can still depend on context, in this case the close proximity of other decaying atoms.

———

Monte Carlo analysis became more feasible with the construction of faster computers, but the method required a good supply of random numbers. In 1947 the newly formed RAND research institute decided to construct a 'random digit generator' based on a radio source producing around 10,000 pulses a second – these were counted electronically, and after each second the final digit of the count was recorded on punched cards.[7] Eventually RAND produced 20,000 cards, each with 50 numbers, for a grand total of 1,000,000 'random' digits.[8]

Of course, RAND conducted many tests to check whether these digits were effectively random, in the sense of satisfying statistical tests that we would expect truly random numbers to pass. They were disheartened to find that a block of 125,000 digits produced on 7 and 8 July 1947, after running the system continuously for a month, had a small excess of odd numbers.[9] They solved this by combining adjacent cards to eliminate any odd–even bias‡ and, after

---

* Quantum computing promises a massive speed-up of this process.

† A half-life of 24,100 years means that 16% of material has decayed by $24,100 \times \log(0.84)/\log(0.5) = 6,060$ years.

‡ Each of the 50 digits on each card was added to the corresponding digit on the previous card, 'modulo 10', which means, for example, that the result of adding $7 + 8 = 15$ was reported as '5'.

further successful checks, the numbers were finally published in 1955. It is a book comprising page after page of digits – perhaps the most tedious book imaginable. But, as I said in the BBC documentary *Tails You Win*,[10] 'Say what you like about it, at least the plot is unpredictable.'

While there is no way of knowing what the next digit will be, there are, however, many predictable patterns. For example, there is a 1 in 100,000 chance of any particular sequence of 5 digits being 12345, and so we would expect 10 such strings in a book of 1,000,000 digits. In fact this sequence occurs 13 times, quite compatible with a Poisson distribution with mean 10. We would also expect one sequence of 7 identical digits: the Poisson approximation says that there is a 37% chance of exactly one, and a 63% chance of at least one such sequence. In fact there is exactly one such sequence, comprising 6666666, as shown in Figure 6.2. This was very satisfying to find, although it would be slightly baffling if you happen to choose this position to start your 'random' set of numbers.

When problems in statistical modelling became too complex for neat mathematics in the late 1980s, statisticians turned to Monte Carlo analysis. A refinement called **Markov Chain Monte Carlo** (MCMC) went beyond just simulating future observations, and also simulated plausible values for unknown quantities assumed to be underlying the data. When combined with the surge in computing technology, this allowed previously impractical analyses to become commonplace. This innovation changed my career, and I spent around fifteen years working on MCMC methods and software. So I owe a lot to Stan Ulam's obsession with card games.

Figure 6.2
Identifying 7 consecutive digits in the book of 1,000,000 random digits: my finger from BBC's *Tails You Win*.

How random are modern random number generators?

Random numbers are now an essential part of modern technology, whether for gaming, simulations or online security, although it might be surprising to learn that most modern random number generators are, in fact, completely deterministic.

You can't just ask a computer to produce a string of random numbers – there has to be an algorithm. Random number generators generally start with a 'seed', say something memorable like 111. They then multiply it by a known big number, add another huge number, remove all but the final digits, and call this a random number. Then the process is repeated, ending up with a sequence of numbers that pass any tests of randomness. And yet the whole process can be precisely reproduced just by knowing the seed number and starting the whole process again – this is extremely useful when wanting to precisely repeat simulations, as we found in our MCMC work.

These algorithms are more accurately known as **pseudo-random number generators**, since they involve no uncertainty whatsoever. But the sequence of numbers is still essentially unpredictable, due to two factors. First, the change from one number to another is highly 'non-linear', in that it does not follow a steady incremental pattern but can follow a wild and unpredictable path. Second, this means they are extraordinarily sensitive to 'initial conditions' – if we chose the root 112, we would get an entirely different and unrelated series. These are the classic properties of a chaotic system, which we shall see more of later.

These ideas lie behind many familiar ways of generating 'randomness', such as flipping coins, throwing symmetric dice, turning roulette wheels or sampling lottery balls from a well-mixed drum. These mechanisms follow the laws of classical physics and so are essentially deterministic, but their extreme complexity makes them unpredictable, and this unpredictability means they will generally be effectively random enough for any practical purposes.

Persi Diaconis, whom we met analysing coincidence in Chapter 4, was a travelling magician who became a professor of probability. He trained himself to be able to flip a coin and to catch it with heads (if that is what he wanted) upwards, and then his team built a coin-tossing machine that could be tuned to make a coin land in a cup as desired.[11] Both are ways of demonstrating that a coin flip is essentially deterministic.*

_____

Anyone who has young children will quickly realize that 'fairness' is an important part of their world, and research shows that by around eight years old many children can grasp that 'picking at random' is one way to ensure such fairness.[12] This is known as *sortition*, and has been used to demonstrate equal opportunity since the selection of juries in ancient Athens.

Of course, it is only fair if the choice is effectively random and the cards are not stacked against you. In 2011 a programming error for the US Green Card lottery resulted in 90% of winners coming from the first two of thirty days allowed for registrations,[13] and it had to be redrawn, which must have led to a lot of disappointment among those who had applied early. Even more notorious was the 1969 lottery for the US draft to the Vietnam War, where very non-random sampling led to 26 out of 31 birthdays in December being drafted, compared to only 14 in January.[14]

Unsurprisingly, as Casanova knew well (Chapter 3), lotteries have always been subject to intense scrutiny about the randomness of the draws, since any suggestion of manipulation would provoke an outcry from disgruntled participants. Which leads to the question

Is the UK lottery truly random?

The main UK Lotto started in 1994 as a 6/49 lottery, in which balls numbered 1 to 49 are mixed in a drum, and then six balls are extracted in a sequence, with their numbers forming the basis for prizes. By October 2015, when there was a change to 59 balls, there had been 2,065 draws. Figure 6.3 shows the distribution of the frequency with which each of the 49 numbers was selected, after 50, 500, 1,000 and 2,065 draws.[15]

After fifty draws there was substantial variability between occurrences:

_____

* Persi can also perform a 'perfect shuffle', in which a pack of cards is split into two parts of 26 cards, and then is riffle-shuffled to produce a perfectly interleaved deck, with alternate cards from each part. To an observer this can look like proper shuffling, but his trick is to do this 8 times in a row, and get back to exactly the original order of the pack. Do not play cards with this man.

ball 39 had only appeared once, while other numbers had appeared eleven times. This inevitably led to claims that 39 was 'due', and indeed after many more draws the distribution of the frequencies got smoother and 39 'caught up'. But, as we saw in Chapter 3, it's important to emphasize that this is not because of some magical compensatory mechanism – between draw 50 and draw 1,000, ball 39 came up 110 times, close to the 116 appearances that would be expected. The crucial insight, as we saw earlier, is that this was sufficient to *swamp* the earlier disparity. The increased smoothness reflects decreasing *relative* variability, even if the *absolute* difference in counts continued to grow, with the maximum final count being 282 draws for number 23, while numbers 13 and 20 lagged sixty-seven draws behind with 215.

Just as RAND did with their million random digits, we can apply statistical methods to check the effective randomness of the lottery draws. The simplest technique is to check whether the distributions in Figure 6.3 are compatible with an underlying **uniform distribution**, representing the assumption that all numbers are equally likely to be drawn.[16] The data pass that test well – the variation between the counts is just about what we would expect.*

While individual draws are effectively random, the pattern of counts is to a large extent predictable. Each number has a probability $6/49$ of being chosen at each draw, and so the count of times a particular number is drawn follows a known binomial distribution.† Figure 6.4 shows the distribution of the 49 total occurrences after 2,065 draws, with 23 at the top and 13 and 20 at the bottom, superimposed with a normal approximation to the binomial. The agreement is reasonable (although the two lowest totals are a little unusual), showing that although we have no idea in advance which the particular 'frequent' and 'infrequent' numbers will be, we can accurately predict the distribution of counts.

This evidence that the numbers chosen are completely unpredictable does not prevent numerous websites giving advice about 'how to win' the lottery. For example, we can read[17] the somewhat contradictory advice that it is better

---

* Technical note: Those familiar with statistical hypothesis testing will recognize that a chi-squared test is appropriate, but some adjustment is necessary due to the fact that a number cannot be chosen more than once in any one lottery draw, and so these are not all independent observations.

† The binomial distribution has mean $2,065 \times 6/49$ and variance $2,065 \times 6/49 \times 43/49 = 221.9$, although the variation in the observed counts would be expected to be slightly reduced by the constraint that the counts must add up to $2,065 \times 6$. In fact, the observed sample variance of the counts is 263.1, somewhat more than the theoretical value, reflecting some rather low counts. This difference is within expected sampling variation from a random process.

both to 'play combinations that are winning the most frequently', while also not to 'play combinations that have been drawn before'. Your choices apparently should have the right ratio of odd to even numbers, and not have 'numbers with the same last digit like 1-11-21-31-41-51' since that 'hasn't happened in the past so you may as well not use it'.

Marginally less useless is the advice not to play consecutive numbers or choose birthdays, since, although it makes no difference to the chance of matching the balls that are drawn, it could affect the number of people you share the jackpot with, if you happen to win. This was clearly demonstrated on only

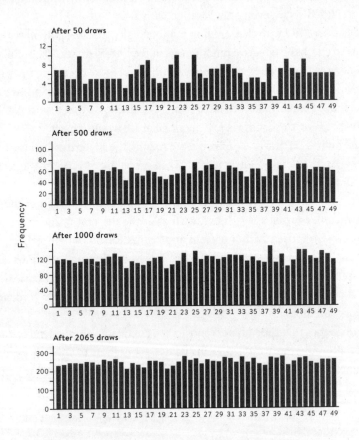

Figure 6.3
Frequency of occurrence of Lotto numbers, starting from the first draw in November 1994 to the final draw in the 6/49 format in October 2015, after which there was a change to 59 balls. The distributions are compatible with what we would expect by chance alone.

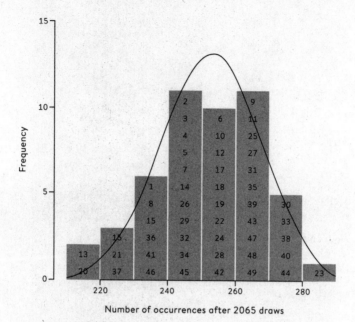

Figure 6.4

The distribution of the number of occurrences of the 49 numbers after 2,065 lottery draws. These follow an approximately normal distribution, as predicted by probability theory.

the ninth draw of the UK lottery on 14 January 1995, when an extraordinary 133 people had to share the jackpot. The way that the forty-nine numbers were arranged on the lottery ticket meant that the balls that were picked (7, 17, 23, 32, 38 and 42) formed a reasonably regular pattern on just two rows, which was an easy set to tick. So scattering numbers irregularly may mean winning more, conditional on the tiny chance of getting the jackpot, which might suggest using the 'lucky dip' facility to obtain random numbers. But frankly it hardly seems worth the effort.

———

Lotteries depend on random draws, but at each draw you need to buy a new ticket. In contrast, UK Premium Bonds is a government-run scheme that offers a monthly draw for prizes, without losing your stake. This is, in effect, a savings system, in which currently around 22 million people, one of whom is me (see Figure 6.5), have in total over £121 billion invested. The draw was launched in 1956 to great fanfare about 'Ernie', the machine that drew the

Figure 6.5
A Premium Bond bought for my fifth birthday in 1958. It has been entered for a prize draw every month since, although I am not sure it has ever won anything, while the £1 stake has devalued to approximately 3% of its 1958 value.

winning numbers, a name contrived from *electronic random number indicating equipment*. Ernie sampled electrical noise as a source of random numbers and had been designed and built at the Post Office Research Station by a team led by Tommy Flowers, who had built the Colossus machine at Bletchley Park to break German codes in the Second World War – arguably the first programmable electronic computer. Similar to the earlier RAND digits, the final numbers were obtained by subtracting outputs from two independent devices, and of course their effective randomness had to be checked. The only woman on the team, Stephanie Shirley,* was given this task, and the numbers passed.[18]

What does random look like?

---

* Shirley is an extraordinary woman. Having come to the UK aged five on a *Kindertransport*, she worked on early computers, got a maths degree in evening classes, and then started an organization for female programmers working from home, signing her business letters 'Steve' to give the impression she was a man. 'Steve' Shirley's organization went on to be a major technology company and sold for £450 million in 2007, with a major shareholding by employees. She has given more than £50 million to charity.

We have notoriously poor intuition about the way pure chance plays out. We see and interpret patterns everywhere, whether it's faces in toast or animals in clouds, lottery numbers that are 'due' or messages hidden in the letters of the Bible. Most of these illusions are harmless, but I have had correspondents who have become deeply disturbed at patterns they detect in everything around them. Indeed, the useful term *apophenia*, meaning the tendency to observe and interpret connections between unrelated things, was invented in 1958 by a psychiatrist, Klaus Conrad, in relation to the early stages of schizophrenia.[19]

I believe the basic problem is that we humans find it very difficult to grasp that *random* does not mean *regular*. A standard trick is to throw a handful of rice over a map and see the clear clusters that emerge – if we were told these represented people with cancer, we would immediately start looking for a reason why one particular area was experiencing so many cases. Whether it is plane crashes or birthdays, randomness is often *clumpy* – while it is a bit simplistic to say that accidents come in threes, we can expect, just by chance alone, that they often do.

Back in the Jurassic days when we had to carry separate devices for phoning, calculating, taking photographs, telling the time, plotting a route, and so on, I used an iPod for playing music. I had about 100 albums, each with 10 songs, for a total of around 1,000 tracks. If I used the 'shuffle' feature to choose the next track at random, I would experience a remarkable number of close matches: after playing 38 tracks there would be a 50% chance of a song repeating itself, while after just 13 tracks there would be a 50% chance of getting another song from the same album.* After complaints from customers, Apple is rumoured to have made the shuffle non-random – in order to make it seem random. Spotify has apparently had to perform the same trick.[20]

I have carried out many classroom exercises with school students, and one of the most successful involves sequences of coin flips. The students sit around tables, and I ask each person to invent twenty coin flips and record a sequence of imaginary heads or tails on a slip of paper which I provide, and then write 'fake' on the back of the paper. Then I lend each one a nice old heavy UK penny and ask them to do twenty real coin flips, again record the results, but this time write 'real' on the back. Then the students on each

---

* We can get these chances from the formulae in Table 4.3. There is a 1 in 1,000 chance of any pair of tracks being the same, and so to have a 50% chance of hearing the same song we need around $1.2 \sqrt{1,000} = 38$ tracks.

table mix up their slips and pass them to the next table, where they have to guess which ones are truly random and which are made up.

The students very quickly get the point of the exercise, as they see that some slips feature quite long runs of heads or tails, while some only have runs of two. I then show them the graphs in Figure 6.6.

The first graph shows that a random sequence of twenty coin flips has a high probability (78%) of a run of at least four heads or tails in a row. This is not intuitive to people, and unless they've done the exercise before, nobody includes such a long run in their made-up sequence – many only feature a maximum run of two, which has only a 2% chance of occurring if the sequence were truly random. Similarly, if we count the number of switches between heads and tails, the average should be 9.5, with most between 8 and 11, but people tend to invent sequences with far more.

It's a fun and popular exercise, and the students can generally correctly

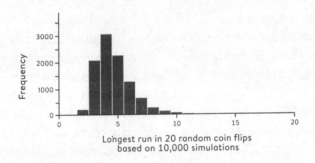

Longest run in 20 random coin flips
based on 10,000 simulations

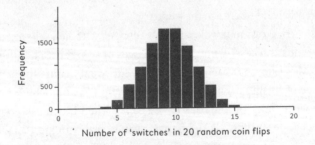

Number of 'switches' in 20 random coin flips

Figure 6.6
Properties of twenty random coin flips, in terms of the longest run of heads or tails, and the number of switches between heads and tails, showing we should expect a run of four or more, and around ten switches. Based on 10,000 simulations.

separate all the slips into true and fake sequences. Hopefully they also learn something about the clumpiness of randomness.

––––––––

Random choice can not only ensure fairness but also ensure that 'winners' and 'losers' are similar, even in ways that we are unaware of. This has numerous scientific applications, for example, 'probability sampling' should mean that those chosen for a survey are representative of the general population. Reliable clinical trials of new medical treatments randomly allocate each volunteer to either receive or not receive the intervention, meaning that the subsequent two groups should be balanced in both known and unknown risk factors. Any subsequent difference in outcomes can then be attributed, up to the play of chance, to the intervention. This simple idea of the randomized clinical trial changed medicine and is responsible for saving millions of lives, as we shall see in Chapter 8.

Randomness can be used in art, as in the works of John Cage and Gerhard Richter, but also to fool an opponent in a game, or even warfare. Your opponent will be desperately trying to work out what your strategy is in order to predict your next move, but adding randomness ensures that what happens next is entirely unpredictable. For example, in Rock, Paper, Scissors, if you use pure randomness to determine your choice, you should be able to defeat any opponent who is trying to guess your choice, although people generally find it very difficult to choose randomly without a device to help.* In contrast, in an analysis of over 11,000 football penalties[21] in which the shooter must decide where to aim the ball, it appears that professional players are indeed able to mimic an unpredictable strategy and often send the goalkeeper diving the wrong way.

Florence Nightingale David, whom we met in Chapter 3 analysing throws of sheep knucklebones, was involved after the Second World War in clearing landmines which had been buried on UK beaches in case of a German invasion. The Germans laid their mines in a methodical, hexagonal pattern, but this made them easier to detect once the first few had been found. With rather more imagination, the British army had used random numbers to decide the spaces between mines, so nobody could detect the pattern. She later reported

––––––––

* In his book *Around the World in 80 Games*, Marcus du Sautoy reports winning at Rock, Paper, Scissors by using the digits of Pi (which are effectively random), to choose the next move.

that 'on the beaches in Norfolk Sands . . . they forgot to record the pattern. It was quite a job. A friend of mine got blown up but not before he had a bright idea. He suggested they take a high-power fire hose and wash the beach. Wash the soil away and expose the mines.'[22]

Random strategies, although mainly effective, may be countered by overwhelming force.

Summary

- 'Pure' randomness can be said to occur when events follow a known probability distribution, unchanged by any external knowledge we may have.
- These 'objective' probabilities may occur at a subatomic level, but in practice most sources produce 'effectively' random events.
- The complexities of the natural world are largely driven by microscopic events, whose causes cannot be determined and can be thought of as 'chance'. Their effects may then be amplified through non-linear 'chaotic' processes.
- Whether the world is truly deterministic or stochastic is not practically important for most analysis, but we need to carefully consider which phenomena we should treat as if they were stochastic.
- Most random number generators are entirely deterministic and contain no randomness at all.
- Physical randomizing devices, such as the way lotteries are drawn, can be effectively random and yet contain predictable patterns.
- Our intuition about randomness is poor – it tends to be far more 'clumpy' than we expect.
- Randomness can be very useful, both for ensuring fairness, representativeness and comparability and deluding opponents.

CHAPTER 7

Being Bayesian

In the UK in June 2021, it was noticed that the majority of people dying from Covid-19 had been fully vaccinated. Should this have been a cause for concern about the vaccines?

There was a lot of uncertainty during the Covid-19 pandemic, and arguments about the effectiveness of social-distancing policies, face masks, and so on will go on for years. Vaccines became a contested issue, both in terms of their effectiveness and their potential harms, and this particular observation caused considerable consternation.

On the face of it, the fact that the majority of Covid deaths were of people who had been fully vaccinated may seem a concerning statistic – were the vaccines actively harmful? But consider the situation in June 2021 in the UK – a vaccine that was claimed to be very effective, but not perfect, at preventing serious illness from Covid had been given to huge numbers of people, with the earliest recipients being the higher-risk groups such as the elderly and clinically vulnerable. So, if we had to make a prediction as to the make-up of people dying from Covid, what would it be reasonable to expect?

Later we shall provide a formal solution to the vaccine question, but you may have already thought of an intuitive answer – the vaccine is not 100% effective at preventing death from Covid, and so if enough people get vaccinated, the 'breakthrough' deaths will outnumber the deaths in the unvaccinated group, even though they were at higher risk. An analogy can help: most people who die in car accidents are wearing seatbelts, but this does not mean that seatbelts are harmful – it's just that nearly everyone is wearing one, and they don't provide perfect protection.

This may seem more a question about statistics than uncertainty, but essentially it is about conditional probability. We know something about the conditional probability of someone dying from Covid *if* they are vaccinated (which is low), but we are interested in the 'inverse' – the conditional probability of someone having been vaccinated *if* they die from Covid (which turns out to be more than ½). A technical solution to this problem involves **Bayes' theorem,** * which we shall see is just a simple consequence of the rules of probability, but has vital ramifications.

I will make the case in this chapter that Bayes' theorem can be considered as a basis for learning from experience, and in principle can form the whole foundation for statistical inference, based on probability theory alone. It has also been claimed that it underlies what happens when humans react to new information – the so-called Bayesian brain. Quite an achievement for a (once obscure) eighteenth-century cleric.

The Reverend Thomas Bayes was born around 1700, educated at Edinburgh University, and became a Presbyterian minister. He later lived in the refined spa town of Tunbridge Wells, where he apparently gave very dull sermons and practised as a skilled amateur mathematician, even being elected a Fellow of the Royal Society. He died in 1761, but his fame rests on a posthumous publication in 1763 of a manuscript found in his papers.[1] This paper, entitled 'An essay towards solving a problem in the doctrine of chances', was submitted by his friend Dr Richard Price, who in his introduction extols the value of Bayes' work on probability, including a claim that it provides an argument for the existence of God.†

Bayes' definition of probability‡ is not exactly transparent but is essen-

---

* A 'theorem' is just a mathematical rule that has been proved to be true under specified assumptions.

† Price claims Bayes' purpose is to 'show what reason we have for believing that there are in the constitution of things fixed laws according to which things happen and that, therefore, the frame of the world must be the effect of the wisdom and power of an intelligent cause; and thus to confirm the argument taken from final causes for the existence of the Deity'. Bayes never made these claims.

‡ 'The probability of any event is the ratio between the value at which an expectation depending on the happening of the event ought to be computed, and the value of the thing expected upon its happening.'

tially the ratio of what you *expect* to win in a bet to what you will get *if* you win. For example, suppose you will get £1 if you win the bet, but beforehand you only expect on average to win 60p – then your probability of winning is $^{60}/_{100}$ = 0.6. So rather than defining expectation in terms of probability, as we did in Chapter 3, Bayes defines probability in terms of expectation. There is no mention of symmetry or long-run frequency – probability is defined entirely in terms of subjective belief. It is perhaps rather ironic, for a Presbyterian clergyman, that his most basic definition should be in terms of gambling.

Bayes is buried in the Nonconformist cemetery in Bunhill Fields in London, where he rests close to luminaries such as Daniel Defoe and William Blake. His work was not properly recognized until the twentieth century, but now the term 'Bayesian' has become standard and he is celebrated in statistics, machine learning and artificial intelligence; Edinburgh University has finally recognized its eminent alumnus by naming its 'Innovation Hub for AI and Data Science' the Bayes Centre, while the Cass Business School, which almost adjoins Bunhill Fields, was renamed the Bayes Business School in 2021 following scrutiny of John Cass's links to slavery.

———————

Bayes' paper followed the style of his sermons in being both long-winded and obscure, but buried in the paper's convoluted language and awful notation[*] is a fundamental idea: initial beliefs about some unknown quantity get revised after observing some data, and the new beliefs subsequently form the basis for estimates and predictions. Before explaining his ideas more formally, we can start with a rather contrived example.

> I am presented with the two identical opaque bags shown in Figure 7.1. I pick a bag at random, pick a ball, note that it is dotted, and put it back. What would be a reasonable probability that I chose Bag 1 or Bag 2? If I draw a ball again from the same bag, what is the probability that it is dotted, and if it is, what is now a reasonable probability for the bags?

———————————————————————————

[*] Bayes used Newton's 'fluxions' notation for calculus, while Pierre-Simon Laplace did so much better when he rediscovered Bayes' work in 1774, and his exposition using Leibniz's notation is immediately recognizable today.

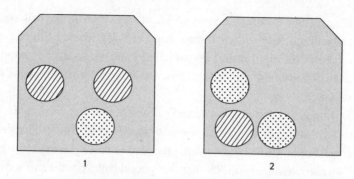

Figure 7.1

Two identical opaque bags, each with three balls; Bag 1 has 1 dotted ball
and 2 striped balls, Bag 2 has 2 dotted balls and 1 striped ball.

Your intuition might be that if I pick a dotted ball, it seems more likely that
I am drawing from Bag 2. This in turn makes it more likely that I will draw a
dotted ball next time. This intuition is right, and Bayes' theorem shows how to
make it precise.

Using the idea of expected frequency trees introduced in Chapter 3, consider what we would expect to happen if we repeated the whole process eighteen times. We would expect to choose each bag nine times, and each of the
six balls three times, as in Figure 7.2. Nine of these balls would be dotted:

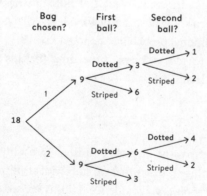

Figure 7.2

What we would expect to happen in 18 repetitions of picking at
random a bag and then a ball from the bag. We see it is dotted, put it back, and draw another ball from the same bag. Nine
of the first balls are dotted, and of these, 5 of the second balls
are dotted.

3 from Bag 1, and 6 from Bag 2. Since we know we drew a dotted ball, reasonable probabilities that we drew it from Bag 1 or Bag 2 would be in ratios 3:6, and so the probabilities for the two bags are ⅓ and ⅔. So after drawing just one dotted ball, we now consider it twice as likely that we are drawing from Bag 2 rather than Bag 1.

Now suppose, having drawn a dotted ball and put it back, we draw another ball from the same bag. Figure 7.2 shows what we would expect to happen on the second draw: $1 + 4 = 5$ of the subsequent draws would be expected to be dotted. So the overall probability that the second ball from the bag is dotted is 5⁄9, which is slightly higher than the probability that the first ball was dotted (½). Our uncertainty has changed as we learn more; the first dotted ball changes our beliefs about which bag we have chosen, which in turn changes our probability that the next ball is dotted.

Out of the 5 possibilities of drawing a second dotted ball, 4 are from Bag 2. So a reasonable probability that I chose Bag 2 is now ⅘ = 80%, showing that after two dotted balls we have rapidly revised our beliefs about the bag in front of us, going from ½ (50%) to ⅔ (67%) to ⅘ (80%).

The process of revising probabilities in the light of experience is a subtle idea and, even if the general principles are intuitive, the mechanics are not immediately obvious. It can help to use some general mathematical notation. Suppose I have a probability for some event A, denoted $\Pr(A)$. We then observe event B, and we want to know how this new evidence changes my probability for A into a new conditional probability, denoted $\Pr(A|B)$.

Bayes' theorem provides a formal procedure for this updating of our beliefs, and takes the basic form

$$\Pr(A|B) = \frac{\Pr(B|A)}{\Pr(B)} \times \Pr(A)$$

This follows directly from the idea of conditional probability in Chapter 3.* The standard terminology is to say that we start with an initial, or **prior probability**, $\Pr(A)$, and after observing evidence B this is revised to a final, or **posterior probability**, $\Pr(A|B)$.

---

* From the definition of conditional probability, $\Pr(A \text{ and } B) = \Pr(A|B) \times \Pr(B)$ and $\Pr(B \text{ and } A) = \Pr(B|A) \times \Pr(A)$. Since $\Pr(A \text{ and } B) = \Pr(B \text{ and } A)$ (the order is irrelevant), this means $\Pr(A|B) \times \Pr(B) = \Pr(B|A) \times \Pr(A)$. Dividing both sides by $\Pr(B)$ gives Bayes' theorem.

We can solve the bags-and-balls problem using Bayes' theorem by letting event A be 'chose Bag 2' and event B be 'picked dotted ball'. Our prior probability $\Pr(\text{chose Bag 2})$ would reasonably be $\frac{1}{2}$, as we have picked the bags at random. After observing a dotted ball, these beliefs are then changed to the posterior probability $\Pr(\text{chose Bag 2} \mid \text{picked dotted ball})$, which by Bayes' theorem is

$$\Pr(\text{chose Bag 2} \mid \text{picked dotted ball}) = \frac{\Pr(\text{picked dotted ball} \mid \text{chose Bag 2})}{\Pr(\text{picked dotted ball})}$$

$$\times \Pr(\text{chose Bag 2}).$$

Now $\Pr(\text{picked dotted ball} \mid \text{chose Bag 2}) = \frac{2}{3}$ from our knowledge of the bags, and $\Pr(\text{picked dotted ball}) = \frac{1}{2}$ as the symmetry of the bags means we are just as likely to choose a dotted as striped ball. Therefore

$$\Pr(\text{chose Bag 2} \mid \text{picked dotted ball}) = \frac{\frac{2}{3}}{\frac{1}{2}} \times \frac{1}{2} = \frac{2}{3},$$

matching the results from the (rather more intuitive) expected frequency tree.

This example illustrates three important points. First, our analysis is based on our assumptions about the *aleatory* probabilities of drawing particular balls, which can be considered as chances, and then by Bayes' theorem are transformed into *epistemic* probabilities, in the sense of personal beliefs about which bag has been chosen. This is a very important step; the observed data, plus our assumptions about how the world works (the play of chance), are converted into judgements about the particular case in front of us.

Second, the repeated draws are done with replacement, and so appear to be physically independent, and yet our probabilities about a dotted ball change. This may at first appear to contradict the idea of independent events. But the draws are only *conditionally* independent, given the (uncertain) choice of bag, and as we've seen, knowledge that a ball is dotted quite reasonably changes our beliefs that the next ball will be dotted.

Conditional independence is a powerful idea that underlies much of statistical modelling, since it is often reasonable to assume that observations are independent *if* we know some common factor that influences them, and so repeated observations allow us to learn about such a common influence.*

---

* Remember from Chapter 3 de Finetti's demonstration that exchangeable sequences can be considered as conditionally independent given some unknown common 'chance' of occurring.

For example, the results from a string of football matches might be assumed to be conditionally independent given the specific teams playing, but can still tell us something about the underlying skill of the teams.

Finally, all this analysis depends on assuming the bags are as claimed, and we are not being lied to. When I do similar exercises with a class, I sometimes substitute a blob of sticky gunge in one bag, which has three purposes; first to try to provoke an entertaining scream, second to teach students that all probabilities are conditional on assumptions, and third that they should be wary of automatically trusting people.

———

On 6 May 2023, King Charles III was crowned at Westminster Abbey in London. Security was intense, and the Metropolitan Police reported[2] that they were using automatic facial recognition to identify whether particular individuals of interest were in the crowds. But how reliable was this system?

A live facial recognition system used by the police is claimed to identify 70% of people on a 'watchlist', while only 1 in 1,000 people generate a false alert. The system picks someone, say 'George', out of a crowd as a match for someone on the watchlist. What is a reasonable probability that George turns out to be truly on the watchlist?

Facial recognition systems can be excellent when used in controlled circumstances using high-quality images – I am very grateful when they enable me to speed through automatic passport barriers. The use of 'live' facial recognition (LFR) to scan crowds to identify individuals on a watchlist is more controversial, not only from the perspective of civil liberties, but also in terms of accuracy, since the images used on the watchlist and the scanning are likely to be of far poorer quality.

Official Guidance from the UK College of Policing[3] uses the following terminology

- The *True Recognition Rate* (TRR) is the proportion of those on the watchlist and who were scanned for whom an alert was then correctly generated. This would be known as the 'sensitivity' in the context of medical screening tests.

- The *False Alert Rate* (FAR) is the number of false alerts generated as a proportion of the total number of subjects processed by the LFR system. Screening tests refer to the 'false-positive rate' or '1 - specificity', which would be the number of false alerts generated as a proportion of the total number of subjects *who were not on the watchlist*, but this is essentially the same as the FAR in this context.

The Metropolitan Police claimed a True Recognition Rate (TRR) of 70%, and a False Alert Rate (FAR) of 1 in 1,000,[4, 5] meaning that only 1 in 1,000 people in the crowd are falsely identified as being on the watchlist.

Suppose that, somewhere among a crowd of 10,000, there were 10 people on the watchlist of interest to the police. Figure 7.3 shows what we would expect to happen to the 10,000 people scanned by the system.

There are 10 individuals of interest, and we would expect the system to identify 7 of these (70% True Recognition Rate). The system would also falsely identify 10 people who were not on the list (1 in 1,000 False Alert Rate). So, in spite of just 1 in 1,000 people being falsely identified, the majority of those identified ($^{10}/_{17}$ = 59%) would be false alerts; the probability that George is actually on the watchlist, having been picked out of the crowd after a match on the system, would reasonably be assessed to be less than ½. This may feel strange and unintuitive, given the apparent accuracy of the system.

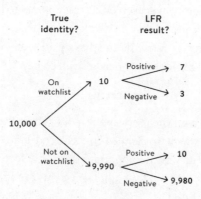

Figure 7.3
What we would expect to happen in a crowd of 10,000 scanned by a live facial recognition system, when there are 10 people in the crowd who feature on a police watchlist, using Metropolitan Police estimates of the accuracy of the system. Of 17 positive identifications, 10 are falsely identified as being on the watchlist.

		True identity?		
		On watchlist	Not on watchlist	Total
LFR result?	Positive	7	10	17
	Negative	3	9,980	9,983
		10	9,990	10,000

Table 7.1
The expected frequency tree of Figure 7.3 displayed as a table.

This analysis can also be displayed in the form shown in Table 7.1. This is another way of viewing Bayes' theorem: the table is constructed 'vertically', using proportions of those on or not on the watchlist, while the quantity of interest – whether the identified individuals are truly on the watchlist – is read off horizontally as a proportion of the live facial recognition results.

If George is incorrectly picked out by the system, perhaps he can be rapidly ruled out from being on the watchlist. But claims of a very low False Alert Rate for the system – for example 1 in 1,000 – may lead the police to have unjustified confidence in the identification; this is known as *base-rate neglect*. This is a special case of a rather unintuitive truth; when the thing you are looking for is rare, then even if a screening test appears to be accurate, the majority of your 'identifications' may be wrong.*

When you are looking for a needle in a haystack, then even if you have good eyesight, most of what looks like needles will be hay.

———

We started this chapter with the story about most people dying from Covid having been vaccinated, which we claimed could be considered as an example of Bayes' theorem. We can now demonstrate this more formally using very rough figures for higher-risk older people, since this group represents the overwhelming number of deaths.

———

* A 2023 study, based on seeding known individuals into crowds, reports that newer software has improved accuracy, with 89% True Recognition Rate and 0.017% (1 in 6,000) False Alert Rate. For our example, this would mean around 11 alerts, 9 of whom would actually be on the watchlist – a considerable improvement.

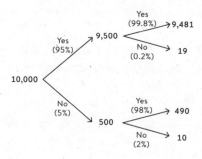

Figure 7.4
Demonstration of why the majority of people who die with Covid
have been fully vaccinated, based on 10,000 older higher-risk
individuals (not using exact numbers). Of 29 people who die with
Covid, 19 have been vaccinated.

Suppose that, of higher-risk older people infected with SARS-CoV2 in June 2021 in the UK, 95% had been vaccinated. Assume the risk of dying among infected higher-risk people who had not been vaccinated was about 2%, and the vaccine was 90% effective against death from Covid, which means the risk in the vaccinated group was reduced to a tenth of its value for unvaccinated, to 0.2%. Figure 7.4 shows what this would mean for 10,000 higher-risk individuals getting infected. We would expect a total of 19 + 10 = 29 Covid deaths, of which the majority, $^{19}/_{29}$ = 66%, had been vaccinated.

Unfortunately, the idea that the vaccines increased the risk of dying from Covid continued to circulate.

While expected frequency trees can help in seeing how Bayes' theorem works, the underlying mathematics is complex. Fortunately, a reformulation of Bayes' theorem not only makes the analysis easier but introduces a quantity that is becoming vital in the criminal justice system – the **likelihood ratio**.

Suppose we are interested in a 'binary' variable which may either be true (denoted A), or not true (not A), and so from the rules of probability, $\Pr(A) = 1 - \Pr(\text{not } A)$. As we saw in Chapter 2, the odds for A is the ratio $\Pr(A)/\Pr(\text{not } A)$ and so, for example, a probability of 0.8 would correspond to odds of $^{0.8}/_{0.2} = 4$.

Bayes' theorem can then be written in what is known as the 'odds form'[*]

$$\frac{Pr(A|B)}{Pr(\text{not } A|B)} = \frac{Pr(B|A)}{Pr(B|\text{ not } A)} \times \frac{Pr(A)}{Pr(\text{not } A)}$$

The **prior odds** for A is $Pr(A)/Pr(\text{not } A)$, and the **posterior odds** is $Pr(A|B)/Pr(\text{not } A|B)$. Bayes' theorem can therefore be written in the simple form

Posterior odds = likelihood ratio × prior odds,

where

$$\text{likelihood ratio} = \frac{Pr(B|A)}{Pr(B|\text{ not } A)}$$

The odds form of Bayes' theorem provides greater insight into what is going on. For example, in the live facial recognition example, Bayes' theorem can be written as

$$\frac{Pr(\text{On watchlist} \mid \text{positive})}{Pr(\text{Not on watchlist} \mid \text{positive})} = \frac{Pr(\text{positive} \mid \text{On watchlist})}{Pr(\text{positive} \mid \text{Not on watchlist})}$$

$$\times \frac{Pr(\text{On watchlist})}{Pr(\text{Not on watchlist})}$$

which becomes

$$\frac{7}{10} = \frac{0.7}{0.001} \times \frac{10}{9,990}$$

revealing the likelihood ratio as $0.7/0.001 = 700$.

For the Covid vaccination question, we can get straight to the likelihood ratio by noting that the vaccine is assumed to reduce the relative risk of death by 90% (its 'effectiveness'), which means a likelihood ratio of

$$\frac{Pr(\text{Covid death} \mid \text{Vaccinated})}{Pr(\text{Covid death} \mid \text{Unvaccinated})} = \frac{1}{10}$$

---

[*] By applying Bayes' theorem to 'not A', we see that $Pr(\text{not } A|B) = \frac{Pr(B|\text{not } A)}{Pr(B)} \times Pr(\text{not } A)$. So when we evaluate $\frac{Pr(A|B)}{Pr(\text{not } A|B)}$ using Bayes' theorem for the numerator and denominator, $Pr(B)$ cancels out and we get the required odds form.

Assuming the prior odds on vaccination of 9,500/500 = 19, the odds form of Bayes' theorem gives us

$$\frac{\text{Pr(Vaccinated | Covid death)}}{\text{Pr(Unvaccinated | Covid death)}} = \frac{\text{Pr(Covid death | Vaccinated)}}{\text{Pr(Covid death | Unvaccinated)}}$$

$$\times \frac{\text{Pr(Vaccinated)}}{\text{Pr(Unvaccinated)}}$$

or

$$\frac{19}{10} = \frac{1}{10} \times 19$$

matching the results from the tree in Figure 7.4.

We shall see in Chapter 10 that likelihood ratios are playing an increasingly important role in summarizing the weight that can be given to forensic evidence. And we have only recently confirmed their vital role in modern history.

How did likelihood ratios help shorten the Second World War?

In a now familiar story, Alan Turing was a brilliant young Cambridge mathematician who led a team at Bletchley Park which broke the Enigma codes and provided essential intelligence for the conduct of the Second World War.* Turing realized that breaking codes was a mixture of analysis and judgement, and Bayesian reasoning was ideal. He described his approach in 1941, first giving the definition we have already seen in Chapter 3:

> The probability of an event on certain evidence is the proportion of cases in which that event may be expected to happen given that evidence.

Which beautifully includes all the ideas that we have seen in this book so far: that all probabilities are conditional on the evidence being considered, that they are personal, and that they can be thought of as an expected proportion of possible outcomes.

---

* I have known two of the Bletchley codebreakers, Donald Michie and Jack Good, but they did not talk about their work. Good wrote a book in 1950 incorporating some of the ideas in this section, but could not discuss applications in codebreaking.

Between 1941 and 1943, Turing relied on assistants (all female) with special paper with holes punched for letters, which could be manually slid along to look for repeated patterns between pairs of messages, which in turn gave suggestions for common settings of the rotors of the Enigma machine. The paper was printed in Banbury, and the whole process was known as *Banburismus*. The aim was to suggest plausible inputs to the electromechanical Bombe computer, and also to rule out other settings in order to reduce the time spent on Bombe runs.*

Turing wanted to use Bayes' theorem to examine the relative probability for competing hypotheses about the underlying settings of the Enigma machine used to generate the coded message. He made two innovations to make the computations feasible; first, rather than repeatedly using Bayes' theorem in its odds form, which requires *multiplying* likelihood ratios, he took logarithms and so reduced the problem to *adding* log(likelihood ratios). Second, he multiplied the log(likelihood ratios) by 10 (later 20), and then rounded the result to a whole number. So the whole process was reduced to adding and subtracting whole numbers, which could be carried out with pencil and paper.†

Turing wrote a paper describing this work, but unfortunately it was not released to the public for seventy years – when it finally appeared in 2012, a modern government codebreaker (known only as 'Richard') said that the material could be released as they had 'squeezed the juice' out of it.[6] Meanwhile, Turing's technique had been rediscovered and become a standard part of machine learning, known as the 'independent Bayes' or 'naive Bayes' classifier, and extensively used in spam detectors and early medical diagnostic systems. The idea of accumulating log(likelihood ratios) was also (independently) developed in the Second World War as the basis for sequential testing of industrial processes, and we used this technique for investigating when mass-murderer Harold Shipman could have been identified (see *The Art of Statistics*).[7]

Like Thomas Bayes, Turing was not widely known in his lifetime, even

---

* Apparently German intelligence did not think Enigma was *theoretically* unbreakable, but that it was *practically* unbreakable due to limitations in computing.

† Log(likelihood ratios) had previously been called 'weights of evidence' by philosopher C. S. Peirce – Turing was apparently unaware of this, but Jack Good later reintroduced this term. The units for rounded '20 × log(likelihood ratios)' were known as 'half-decibans', and Good claims this is about the smallest change in weight of evidence that is directly perceptible to human intuition.

though the work of the Bletchley codebreakers is said to have shortened the war by between two and four years[8] and certainly saved vast numbers of lives. In 1952 he was prosecuted for 'gross indecency' with another man, and when he died in 1954 the inquest recorded his death as suicide after apparently eating from a poisoned apple. He was finally pardoned by royal prerogative in 2013 and is now, like Bayes, widely honoured – the Alan Turing Institute is the UK's national institute for data science and artificial intelligence, and as of 2021 he appears on the £50 banknote. But his insights into Bayesian reasoning remain largely uncelebrated.

———

The preceding stories have shown how Bayesian thinking enables us to update our beliefs on the basis of multiple pieces of evidence. So far, we have restricted ourselves to beliefs about propositions that are either right or wrong, but it is natural to extend this process to learning about any underlying but currently unknown quantity in the world, for example the true population of a country or the average effect of a drug. This takes us, of course, to the ideas of statistical inference, which, although not the topic of this book, is inseparable from any discussion of uncertainty.

There are different schools of thought about statistical inference (see Chapter 8). Put very briefly, the Bayesian approach says that we have a prior probability distribution about some unknown state of the world, we observe some relevant data, and then our prior is updated to a posterior probability distribution by Bayes' theorem.* That's it! Of course, in practice there are numerous complications related to appropriate assumptions about the process that gave rise to the data, and the challenge of actually computing the answer meant that, realistically, complex examples were not feasible until the late 1980s. But the crucial insight is that no other principles are required – the whole of statistical inference can be reduced to probability theory, which is why it features in this chapter.

———

* Unknown characteristics of the world are represented by parameters of models, and generally given Greek letters such as θ (theta). We assume our personal uncertainty about θ can be expressed as a prior distribution $Pr(\theta)$. We then observe some data x, and we assume a probability model $Pr(x|\theta)$ that expresses how the probability of observing x depends on θ – this measures the support provided by x for different values of θ, and is technically known as the likelihood (hence the likelihood ratio is preceding sections). Then Bayes' theorem shows how these two elements should be combined to form a posterior distribution $Pr(\theta|x)$ which summarizes our revised beliefs about θ having observed x, where $Pr(\theta|x) = Pr(x|\theta)Pr(\theta)/Pr(x)$. This may sound simple, but there have been decades of work on efficient ways to calculate posterior distributions, particularly when θ has many dimensions and x is huge.

Bayesian inference has long been mired in controversy, since it rests on an acknowledgement that probability is a quantification of personal uncertainty about some currently unknown fact, and that the prior and posterior probabilities do not exist as properties of the outside world but are constructions based on currently held assumptions. These ideas run through the whole of this book, and so should be familiar by now, but they can be extremely challenging to those brought up to believe that probability is defined as a long-run frequency of some repeatable event and that statistical inference should be 'objective'.

———

We have repeatedly celebrated those like Richard Feynman who have the humility to admit they don't know and are ready to change their minds when confronted with surprising evidence. But while this may be an admirable characteristic of humans, is it possible for it to be built into automatic learning systems? In other words

How can we express humility in mathematics?

Remember the two-bags example in Figure 7.1, where Bag 1 has 1 dotted ball and 2 striped balls, and Bag 2 has 2 dotted balls and 1 striped ball. Suppose we pick a bag at random, and then draw a sequence of balls from that bag, replacing each after it is drawn, and after each draw assess the probability that we have picked Bag 1 or Bag 2, and the probability that the next ball will be dotted. Suppose we have actually picked Bag 2, the one with more dotted balls. Figure 7.5(a) shows a simulation of what might happen as we keep on repeating the process of drawing balls and putting them back. The probability that we have chosen Bag 2 wobbles around a bit but then steadily tends towards 1, while the predictive probability that the next ball is dotted goes towards $2/3$. Just what we would expect, as we steadily get more confident in the truth.

But what if we have been fooled? Suppose the person who supplied the bags has lied, and in fact both bags contain three dotted balls. Then all the balls we draw will be dotted, but we will keep on updating our beliefs using Bayes' theorem, oblivious of the trick that has been played. Figure 7.5(b) shows what will happen – our probability that we have chosen Bag 2 will tend

(a) Balls taken from Bag 2 (2/3 dotted)

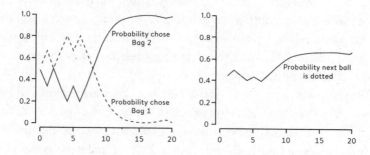

(b) Balls taken from 'fake' trick bag (3/3 dotted)

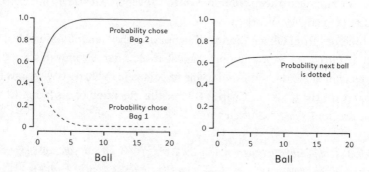

(c) Balls taken from 'trick' bag, given small prior probability

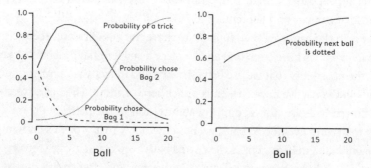

Figure 7.5
We are told that Bag 1 contains 1 dotted ball and 2 striped balls, Bag 2 has 2 dotted and 1 striped ball. We pick a bag at random, pick a ball, put it back, and assess our probability about which bag we have picked (*left-hand graphs*) and the predictive probability that the next ball is dotted (*right-hand graphs*). (a) a simulated example of when balls are drawn from Bag 2; (b) when the balls are drawn from a 'trick' bag with 3 dotted balls, but our model for the situation doesn't allow for that possibility; (c) when balls are drawn from the 'trick' bag, but a small prior probability has been given to this eventuality.

to 1, rather faster than before as we will only draw out dotted balls, and the predictive probability that the next ball is dotted will again go towards $\frac{2}{3}$, since that is the only option available, but is incorrect – the true probability is 1. Of course, at some point we will get suspicious at the unrelenting stream of dotted balls and demand to check the bags, but not before Bayes' theorem has produced some inaccurate predictions.*

We can, however, easily avoid this behaviour if we are prepared to follow **Cromwell's Rule**. This refers to a principle, popularized by prominent Bayesian statistician Dennis Lindley, that you should not give probabilities of 1 to any event unless it is logically true, such as 2 + 2 = 4, and never give probability 0 to any event unless it can be logically shown to be false – in other words, have the humility to keep our minds open to events that we had not expected, and so be ready for surprises.

It comes from Oliver Cromwell's appeal to the General Assembly of the Kirk (Church) of Scotland on 3 August 1650, when Cromwell's army was camped outside Edinburgh and trying to persuade the Kirk to withdraw their support for the return of Charles II after the execution of his father Charles I the previous year. Cromwell wrote

> *Is it therefore infallibly agreeable to the Word of God, all that you say? I beseech you, in the bowels of Christ, think it possible that you may be mistaken.*[9]

This appeal was ignored, and Cromwell soundly defeated the Scots at the Battle of Dunbar on 3 September 1650.†

So, in the bags example, suppose we have the humility, and scepticism, to doubt what we have been told and so place a small initial probability, say 1%, on the possibility that we are being tricked and in fact the number of dotted balls in the bag we have chosen is either zero or three. Then as a steady run of dotted balls are drawn, our probabilities follow the paths in Figure 7.5(c). Up to around five dotted balls, we broadly follow that previous path in placing our belief in Bag 2. But as the dotted balls keep coming, our sceptical view that we might be being tricked starts increasing, and after twelve dotted balls

---

* There are Bayesian techniques for detecting when data do not fit the assumptions, ideal for situations such as this.

† Carlyle reports some comments about the cleanliness of the impoverished Scots: 'they are much enslaved to their Lords, poor creatures; almost destitute of private capital – and ignorant of soap to a terrible extent!'

we rapidly come to the conclusion that we have been conned and there are in fact three dotted balls in the bag. Our probability that the next ball will be dotted appropriately heads towards 1.

Cromwell tells us to retain a smidgeon of scepticism about what we may take for granted, such as the honesty of the person sorting out the bags of balls. We are essentially dealing with one type of 'unknown unknown', comprising inappropriate assumptions that we make without thinking, and essentially turning them into 'known unknowns'. And going back to our discussion of hedgehogs and foxes in Chapter 2, we see that the answer to Cromwell's exhortation is to act like a fox, be ready to be surprised, and to be humble and flexible enough to admit that things have changed. And, rather remarkably, all this can be expressed mathematically in terms of prior distributions, by just allowing a small probability that 'we may be mistaken'.

Of course, it is not possible to put a small prior probability on every eventuality, and some surprises might just make us throw away our initial assumptions and completely rethink our ideas. Such humility should prove useful in all aspects of life – not just statistical inference. And if some people have an ability to change their minds dramatically when solid evidence comes along that conflicts with their previous views, perhaps humans really do operate in a Bayesian way.

———

Our brains, and our consciousness, do not experience the world directly. All our perceptions are filtered through the senses, whether sight, touch, hearing, and so on, which send neurological signals to the grey mush in our heads – our experience would not be very different if our brains were next to our liver. Every moment there is new information to process and a response in our mind and body is generated. But all this new data streaming into our brains is (in general) not radically different from what we have sensed before, and so we have strong expectations of what we are going to experience in the next instant.

This seems self-evident, but it leads naturally to the idea of the 'Bayesian brain'.[10] We have an internal 'mental model' for how the world works, constructed from all our experiences throughout our lives. When combined with what we are sensing in the current situation, this model generates an expectation about what will happen next – this can be considered a prior distribution. We then observe some evidence from the outside world, and then our beliefs about what is going on around us are revised using at least an approximate form of Bayes' theorem, in order to try to minimize the gap between what we expect and what we observe. Autonomous road vehicles operate in this way, using explicitly Bayesian updating algorithms.

The simple example of the dotted balls in bags illustrates how the brain might function in a Bayesian way. If the prior expectations are rigid, then we know we can try to force everything into that framework, even if the data say otherwise – just think how easily we construct reasons for a usually reliable person being late. But if we entertain even a small doubt about their reliability, then we might rapidly shift to believing that they have simply forgotten the meeting.

Although the concept of a Bayesian brain appears to explain processes such as learning, reasoning and perception, it is still not established how closely the neurological changes actually match those dictated by Bayes' theorem. But the crucial lessons are that (a) we are constantly updating our uncertain beliefs about the world, and (b) the only way we can do this is by having some internal model for how the world works.

These internal models are implicit and only indirectly revealed by our perceptions, beliefs and actions – in contrast, later in the book we shall come to the challenges of constructing explicit mathematical models of the world which are open to full scrutiny. It may seem dehumanizing to think of us as large statistical inference engines, but if we acknowledge that our comprehension is based on a model rather than on direct experience of reality, then perhaps we might be more understanding, and perhaps more sceptical, of attempts to represent the complexities of reality in equations.

––––––––––

We've come a long way in this chapter, starting with a question about why most Covid deaths are among the fully vaccinated, through the powerful idea of summarizing the weight of evidence through likelihood ratios, and ending in a theory of human perception and cognition. Some of this has been unavoidably technical, but I hope the basic principles have come through. In summary, if we accept that uncertainty is part of our personal relationship with the outside world, then Bayes provides a model for how we should change our beliefs in response to our constantly changing experience.

Summary

- Bayes' theorem comes from the basic rules of probability and shows how our beliefs should change in response to new evidence.
- It can illuminate some unintuitive phenomena, such as screening systems having apparently high 'accuracy' and yet the majority of claims of positive identification being wrong.
- When comparing alternative propositions, the evidential support from a piece of information is summarized by the likelihood ratio.
- Used naively, Bayes' theorem can be slow to adapt to unexpected changes. But having some humility, and expressing just a little doubt about the correctness of assumptions, allows a rapid shift to a revised set of beliefs.
- Our brains work in a Bayesian way, having prior expectations that are revised in the light of sensory inputs.

CHAPTER 8

Science and Uncertainty

'Dans les champs de l'observation le hazard ne favorise que les esprits
   préparés.'
('In the field of observation, chance favours only the prepared mind.')

—Louis Pasteur, 1854

If our view of science is based on how we were taught in school, or how it is portrayed in the media, we may think of it as a body of confirmed laws and facts about how the world works. This *settled* science is vitally important, and good enough for most of us. But it is not the concern of active scientists, who are more focused on conducting research to push out the boundaries of our knowledge. And just like when exploring a physical frontier, this effort is characterized by uncertainty.

Using the language introduced in Chapter 1, we can identify different 'objects' that scientists, in the broadest sense, may be uncertain about. These might include –

- The magnitude of physical quantities: for example the speed of light, and the distance to stars.
- How many things: the number of tigers in India, and the number of migrants entering the UK each year.
- 'Virtual' quantities that cannot be directly observed and need to be inferred: Gross Domestic Product (GDP), the average effect of a pharmaceutical drug, or mean global change in temperature over the last century.
- What has happened in the past: the process of evolution on earth.

- What exists: life on other planets, and the site of lithium deposits on earth.
- The fundamental nature of our universe: the role of dark matter, and the existence of subatomic particles such as the Higgs boson.

Note that this list only concerns *epistemic* uncertainty about what has happened, what is going on at the moment, or how the world works – we shall deal with the even trickier problem of predicting the future in Chapter 11. Warning: even with these limitations, this chapter is still quite challenging, but it covers some of the most important material in the book.

Of course, if we could directly and precisely observe things, whether a quantity or a fact, then we wouldn't need to bother with uncertainty – we would just be able to say what was the case. But we can rarely do this, and we are left with making observations that are directly or indirectly related to what we are interested in, and then drawing conclusions based on the evidence provided by that data. And that data will display variability, some of it unexplained. Statistical inference is the process of turning that variability into an assessment of uncertainty about the object of interest.*

When we consider statistical approaches for characterizing uncertainty about quantities, facts or scientific hypotheses, this will inevitably mean introducing the traditional ideas of measurement error, confidence intervals, P-values, and all those other concepts that you may have struggled with on statistics courses. Often overlooked is the danger that our conclusions may be overly sensitive to questionable assumptions embodied in the statistical model for how our data occurred. Once this challenge to the calculated uncertainties is acknowledged, researchers may add quantified judgement, make the models even more elaborate, conduct extensive sensitivity analysis, or combine outputs from multiple models. In the end, it may not even be appropriate to put all our uncertainty into numbers.

Many of the scientific issues I listed at the start of the chapter are contested and occasionally subject to fierce debate, and audiences may not realize just how much is unknown. Acknowledging uncertainty might be uncomfortable for scientists as, in everyday language, when we say we are uncertain it could suggest that we haven't much of a clue. But this discomfort needs to be overcome, as science has a natural language of uncertainty that can convey both what is known and unknown, and the

---

* Shameless plug: see *The Art of Statistics* for a fuller discussion of statistical science.

appropriate confidence in any conclusions. We need to proclaim our uncertainty with pride.

We start with an area of apparently 'hard' science which, perhaps paradoxically, fully embraces quantified judgement.

————

The science of measurement is called *metrology* and dates back to the French Revolution, when there was a political imperative to standardize units across France. And so the metre, kilogram and litre were born. The International Bureau of Weights and Measures is still based in France and known by its French title BIPM (Bureau International des Poids et Mesures), and produces the bible of measurement – the Guide to the Expression of Uncertainty in Measurement (commonly known as the GUM).[1]

The GUM distinguishes two types of evaluation of uncertainty:

- *Type A*: 'by the statistical analysis of series of observations', in other words the standard model-based calculations embodied in computer packages.
- *Type B*: 'by means other than the statistical analysis of series of observations . . . evaluated by scientific judgement based on all of the available information on the possible variability'. Type B uncertainty is expressed as a 'subjective probability' distribution representing 'degree of belief'.

Each type of uncertainty should be summarized by a probability distribution, and then the two types combined by standard techniques.* The US National Institute of Standards and Technology[2] takes a similar approach, agreeing that Type B uncertainty 'is usually based on scientific judgment using all the relevant information available'.

It may seem surprising that the bodies responsible for weights and measures, which we might think of as the most objective of scientific processes, explicitly recommend subjective assessment of uncertainty. But this reinforces the central message of this book – that uncertainty is a personal relationship with the world, and judgement is inevitable. The GUM recommends doing everything possible to quantify uncertainty using statistical procedures, but then to add on a judgement of additional uncertainty that has not been captured by data

———

* For example, if the calculated Type A uncertainty is summarized by a variance v_A, and the additional Type B uncertainty is judged to have variance v_B, then the overall uncertainty is expressed as a total variance $v_A + v_B$.

analysis. We shall find this vital and fundamental idea repeated throughout this chapter.

Modern statistical science, rather oddly, has placed little emphasis on the basic problem of measurement, but throughout history people have struggled to get accurate estimates of lengths, speeds and weights. The standard process is to make a number of independent measurements, exerting every effort both to eliminate biases and reduce unnecessary variability, and then take some sort of average, often the mean. This enables researchers to answer questions such as

How certain are we about the speed of light?

In 1879, 27-year-old Albert Michelson constructed an ingenious apparatus to estimate the speed of light in a vacuum, a quantity that is generally denoted c, based on light reflecting on to a spinning mirror. He estimated c to be 299,944.3km/sec, and a 'Type A' margin of error of +/- 15.5 can be calculated from his measurements. But Michelson reported a much wider interval, +/- 51, taking into account his judgement of the systematic bias in his apparatus. So years before the recommendations in GUM, Michelson was assessing his Type B uncertainty.

Since 1983 the speed of light has been set as 299,792.458km/sec,* which is 152km/sec less than Michelson's estimate. Michelson's claimed margin of error was therefore considerably too small – his results were fairly *precise*, in that they were fairly closely bunched, but not so *accurate*, in that they systematically overestimated the true value. Yet he still managed to get within 0.05% of the final accepted value, which was an extraordinary achievement for that time.

As Max Henrion and Baruch Fischhoff showed in 1986, similar optimistic claims about margins of error are typical of the history of estimating physical constants. Figure 8.1 shows how the official recommended values for the velocity of light c changed between 1929 and 1973, and compares them with the current accepted value.[3]

Lower estimates of c in the 1930s and early 1940s led some physicists to

---

* The metre is now defined as the distance that light travels in 1/299,792,458 of a second, and so the metre and the speed of light are pegged together for the foreseeable future.

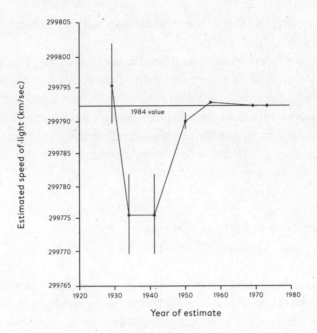

Figure 8.1

Recommended values for the velocity of light in a vacuum, 1929–73, showing the claimed margins of error were over-optimistic.[4]

suggest that the speed of light was actually slowing down, although Raymond Birge, head of the physics department at the University of California Berkeley, declared in 1941, 'Thus, after a long and, at times, hectic history, the value for *c* has at last settled down into a fairly satisfactory 'steady' state. The same thing can now be said of other important constants.' He spoke too soon, as just nine years later the estimate of *c* had changed dramatically. He was wrong about the other constants too, as Henrion and Fischhoff have shown that recommended values for Planck's constant, the charge of an electron, the mass of an electron, and Avogadro's number have all changed since 1941 to values far outside their previously claimed margins of error.

So why are the claimed uncertainties too small? The crucial insight is that these margins of error are calculated assuming that a list of assumptions are all correct. But there are five basic assumptions in the determinations of physical constants that are open to question:

1. *No systematic bias*: we need to assume that, were a huge number of observations made, their mean would eventually tend towards the true

value, with no systematic under- or overestimation. This depends on the skill and insight of the experimenter, and Michelson should be congratulated for his accuracy.

2. *Accurate estimate of variability*: the spread of the observations should genuinely reflect the precision and reliability of the measuring apparatus, which is assumed the same for all data points.

3. *Independent observations*: if the observations are related to each other, for example by some carry-over of a disturbance, then the 'effective' number of observations is lower than claimed.

4. *The mean has an approximately normal distribution*: this is perhaps the least questionable assumption, as it is guaranteed by the **central limit theorem** for a wide range of underlying sampling distributions.

5. *The data have been reliably reported.*

A notorious example of underestimating variability (breaking Assumption 2) arose from Robert Millikan's famous 1912 'oil-drop' experiments to measure the charge on an electron. Despite his claiming that 'this is not a selected group of drops but represents all of the drops experimented on during 60 consecutive days', later examination of his notebooks revealed that he had excluded results that he felt were unreasonably discrepant: Franklin[5] reports that of 107 observations made after his apparatus was stable, Millikan rejected 49 of them because, for example, one was 'too high by $1\frac{1}{2}$'. This suggests Assumption 5 is also unjustified.

It turned out that Millikan 'trimmed' both high and low results, which meant that he did not give an overall bias to the results, but did make the claimed variation too small, and this in turn led to a reduction in the claimed uncertainty. Millikan was awarded a Nobel Prize in 1923, but the ethics of his actions continue to be debated. It's been suggested that he committed fraud, but the undoubted accuracy of his conclusions is undisputed; he got within 1% of the currently accepted value (although he claimed to be within 0.2%).

———

We've already referred to the idea of a **statistical model**, which is an attempt to capture the important and relevant features of reality in a mathematical form. Such a model embodies assumptions about how the observed data relate to underlying quantities, generally called *parameters* (and traditionally given Greek letters), which are intended to correspond to aspects of reality that are of interest, such as the average effect of a drug. Statistical inference is the process of both estimating parameters and assessing the uncertainty around those estimates.

The following example features a parameter of considerable importance.

> What is the effect of the steroid dexamethasone on the survival of patients who are severely ill with Covid-19?

Soon after the start of the SARS-Cov-2 epidemic in the UK, the RECOVERY trial started to test therapies in people hospitalized with Covid-19. It was a 'platform' trial, meaning that it consisted of a series of overlapping studies, where each patient might be entered into a number of simultaneous trials. In one experiment, conducted between 19 March (before the start of the UK lockdown) and 8 June 2020, 6,425 patients were randomly allocated to receive either dexamethasone, an inexpensive type of steroid called a glucocorticoid, or to receive the usual care as a 'control': the randomization was in proportion 1:2, so that roughly twice as many patients were in the control group.[6]

A variety of outcome measures were recorded, but we focus on 28-day survival in the most severely ill group of patients, who were receiving mechanical ventilation at the time of randomization. Table 8.1 shows the outcomes of the 324 patients randomized to receive dexamethasone, compared to the 683 randomized to usual care.

The observed **relative risk** was 0.71, with a '95% **confidence interval**' (explained below) ranging from 0.58 to 0.86; subtracting these numbers from 1 means the 28-day mortality rate was 29% lower in the group randomized to dexamethasone, with a 95% confidence interval for this relative risk reduction ranging from a 14% reduction to a 42% reduction. This shows considerable uncertainty, in spite of the large number of patients randomized. The absolute risk difference, shown in the last row of the table, was observed to be −12%, meaning that of eight people randomized to receive dexamethasone, one (12% of the eight) would survive twenty-eight days who would otherwise not have.

All these estimates and confidence intervals are calculated according to standard formulae and can be generated in a fraction of a second using (reasonably) friendly software. Thousands of such analyses are carried out each

Quantity to be estimated	Number randomized	Number who had died by 28 days post-randomization	Estimate of true underlying quantity	95% confidence interval
Risk in group randomized to receive dexamethasone (treatment group)	324	95	29.3%	24.4% to 34.6%
Risk in group randomized to receive usual care (control group)	683	283	41.4%	37.7% to 45.2%
Relative risk			29.3/41.4 = 0.71	0.58 to 0.86
Relative risk reduction			0.29	0.14 to 0.42
Absolute risk difference			29.3% – 41.4% = –12.1%	–5.7% to –18.5%

Table 8.1

Comparisons of 28-day mortality in those randomized to receive or not to receive dexamethasone, for patients on mechanical ventilation at the time of randomization. Those randomized to dexamethasone showed a substantial improvement in 28-day survival. The 'relative risk' is the risk in the group randomized to the new treatment divided by the risk in the control group.

day, often using large and complex datasets, and the results published in academic papers and government reports. It has become completely routine.

But what does it all actually mean? The confidence intervals apparently express uncertainty about the estimates, but their technical definition is somewhat tortuous. Essentially, if we repeatedly calculate such intervals in study after study, and *if the assumptions of all the statistical models we use are correct*, then 95% of the intervals will contain the true value. According to this formal definition, we cannot make any statement about the probability that this particular interval contains the true value, just the long-run properties of using this procedure. Unsurprisingly, people have problems with this complex and unintuitive definition, and often say something like 'we can be 95% confident that the true value lies in the interval'.

Furthermore, in all the vast number of analyses carried out, there is no mention in the computer output that the intervals are, as we said, only precisely valid if *all* the assumptions of the model are correct. For example, the assumptions underlying the analysis featured in Table 8.1 include

1. The observations are independent, for example there is no factor that might make patients treated closer in time have more similar outcomes.
2. All the patients in each group have the same probability of surviving twenty-eight days.
3. All the patient data is reliably recorded.

These assumptions define the statistical model that the number of deaths at twenty-eight days follows a binomial distribution (see Chapter 3) within each group.*

Unfortunately, the assumptions listed above are not all true. First, the observations are not completely independent, since there are bound to be common factors that influence the care of patients that are close to each other in space and time, whether it's the hospital in which they are being treated or changing care regimes. Second, patients will differ in their risk for all sorts of reasons. In contrast, the third assumption appears reasonable in this example, as presumably we can have confidence in the reliability of the data in such a well-organized and meticulous trial.

But just because the underlying assumptions are not strictly true, it does not mean that the analysis is fundamentally flawed. In this case the signal is so strong that, for example, a model that allows the underlying risk to vary between patients will make little difference to the overall conclusions. It would be different if the results were marginal, when it would be appropriate to do extensive sensitivity analysis to alternative assumptions and acknowledge contributions to any volatility in both the estimate and the scientific conclusions.

Crucially, since the patients have been randomized, the two groups should be balanced, not only for factors that we know can influence outcomes, such

---

* The published paper adopted a more sophisticated model using data on the exact day of death, and estimated an age-adjusted **hazard ratio** – the relative risk reduction of dying each day in the dexamethasone group compared to control. This was 0.64 (95% confidence interval 0.51 to 0.81), meaning that, within each of three age groups (less than 70, 70 to 79, and greater than 80 years old), on average the daily risk of death was estimated to be 36% lower in the dexamethasone than the control group. Almost exactly the same conclusion can be reached using the 28-day survival data alone.

as disease severity, but also with respect to factors that we are unaware may be important. Any observed differences in the outcomes, allowing for chance variation, are therefore due to the randomized group. We can therefore conclude causality rather than just correlation.

————

It is rather humbling to acknowledge that every published statistical analysis depends on a model that incorporates numerous assumptions which are either demonstrably false or cannot be confirmed. This observation led British statistician George Box to his much-quoted aphorism

All models are wrong, but some are useful.

which neatly summarizes the accumulated wisdom from a lifetime of statistical analysis. Models are mathematical representations of reality – they are the map, not the territory. In her book *Escape from Model Land*,[7] Erica Thompson suggests thinking of models as metaphors, even caricatures, of the world – a good model will incorporate the essential features, but not be concerned with unimportant detail. George Box went on to say, 'Since all models are wrong the scientist must be alert to what is importantly wrong. It is inappropriate to be concerned about safety from mice when there are tigers abroad.'[8] So the challenge is not to determine which model is 'correct' – this is not a meaningful aim as a correct model does not exist – but which is adequate for whatever purpose is in mind, whether it be explanation or prediction.

Unfortunately, rather than the flexible exploratory approach to statistical modelling recommended by Box and others, a rigid attention to 'statistical significance' has come to dominate much of scientific publishing. This is what we must turn to next.

P-values, significance tests and uncertainty

Following widespread statistical practice, we can also calculate a **P-value** for the observed difference between the groups in the dexamethasone study. This is the calculated probability of observing such an extreme statistic, assuming the hypothesis that there is actually no underlying difference in the risk in the two randomized groups, and the observed effect was solely due to the play of chance – this is known as the **null hypothesis** of 'no difference'. The calculated

P-value in the dexamethasone example is P = 0.0003, which is very small, meaning it was very unlikely that such a big difference would be observed, if all that had been operating was the play of chance. With such a small P-value, the standard practice would be to reject the null hypothesis and declare the results 'statistically significant'.

There is, however, increasing disquiet in the scientific community with this traditional process.[9] The many reasons for concern include:

1. The use of arbitrary thresholds to declare results 'significant', such as P < 0.05, leads to an inappropriate tendency to dichotomize findings into a 'discovery' or not. In particular, 'non-significant' results are often wrongly interpreted as meaning 'no effect'. As statistician Andrew Gelman has said, 'it seems to me that statistics is often sold as a sort of alchemy that transmutes randomness into certainty, an "uncertainty laundering" that begins with data and concludes with success as measured by statistical significance.'[10]

2. The P-value is not a measure of uncertainty about the null hypothesis, and it is certainly not the probability that the null hypothesis is true. Rather, it is a measure of the compatibility of the observed data with the null hypothesis.

3. Carrying out multiple significance tests hugely increases the probability of an incorrect 'significant' finding somewhere.

4. Just as for the confidence interval, the calculation of the P-value depends on *all* the assumptions in the statistical model being met.

5. The null hypothesis is not even plausible, as we would never expect exactly zero effect of a treatment, and so is a 'straw man' which, with sufficient data, will always be rejected.

But the use of significance testing continues regardless. While it can provide a broad impression of the extent to which the data are compatible with specific scientific claims, the big problem is the obsessive attention to specific thresholds such as P < 0.05, and the consequent misinterpretation of the results.

As we saw previously, the formal definition of a 95% confidence interval can be rather baffling. But there is another interpretation which might bring some insight; it is the range of possible null hypotheses that cannot be rejected in a significance test with a P-value less than 0.05. A prominent epidemiologist, Sander Greenland, has therefore proposed replacing the label 'confidence interval' with the term *compatibility interval*, emphasizing that the interval contains values of the underlying parameter that are compatible with the observed data under the assumed statistical model.[11] This all seems rather sensible, although it is unclear whether it will catch on.

———————

Some researchers (including me) have suggested that a Bayesian approach might circumvent some of these challenges. As we saw in Chapter 7, this requires specification of a prior probability distribution for the parameters of interest – this is then combined using Bayes' theorem with the 'likelihood', which summarizes the relative support for different values of the parameters provided by the data. The resulting posterior distribution encapsulates a judgement about the true value of the parameters. As I said before, the whole process is 'just' probability theory.

In the dexamethasone example, the basic unknown parameters are the underlying mortality risks in the treatment and control groups. Suppose we give each of these a 'uniform' prior distribution, which is essentially saying that before we observe any data we think any value between 0 and 100% is equally likely. This may seem implausible, but the authors reported that at the start of the Covid-19 pandemic they had so little idea of what the mortality rates might be that they could not carry out the standard calculations to decide the sample size of the trial, since that requires some judgement about the likely underlying risks. And, in any case, in this example the data overwhelm the prior and so its exact form is not important.

These prior distributions can then be combined with the binomial likelihood from the data, to produce the posterior distributions shown in Figure 8.2(a).* There is clear separation between the two groups, meaning we should be confident there is a true difference.

It is impossible to derive a 'nice' mathematical form for the posterior distribution of the relative risk or absolute risk difference, but it is straightforward to carry out a Monte Carlo analysis; by simulating 100,000 pairs of values from the posterior distributions in (a), and then calculating the ratio and difference between each pair, we can obtain the distributions shown in (b) and (c). The uncertainty in the risk ratio and the difference is clearly displayed, and we can assess probabilities of various events, such as there being a probability of

———————

* A uniform prior combined with a binomial likelihood based on r events out of n opportunities produces a posterior distribution following a **beta distribution** with parameters $r+1$ and $n+1$.

(a)

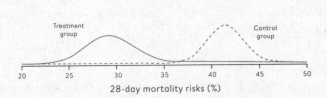

28-day mortality risks (%)

(b)

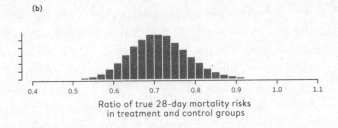

Ratio of true 28-day mortality risks
in treatment and control groups

(c)

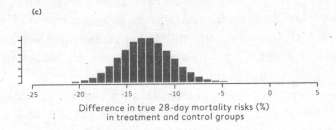

Difference in true 28-day mortality risks (%)
in treatment and control groups

Figure 8.2
(a): Bayesian posterior distributions for underlying 28-day mortality risks in the two randomized groups; (b) the relative risk; and (c)
absolute risk difference. (b) and (c) are based on 100,000 simulated
values from the posterior distributions in (a).

around 99.985% for the treatment group having a *lower* underlying mortality rate than the control group, and 17% probability for the treatment group
having *more than 15% lower* mortality than the control.

Both traditional and Bayesian approaches come up with similar conclusions, although I personally prefer the Bayesian analysis since

- The posterior distributions provide a direct visualization of the support
  for different values of the unknown quantities.
- We do not need to introduce the idea of a null hypothesis.

- We can avoid P-values by directly assessing a probability for events of interest, such as whether the risk difference in favour of the treatment group is larger than 15%.

Nevertheless the two approaches still have much in common, since each assumes a statistical model in which the observed outcomes have a binomial distribution with a common risk of dying before thirty days, and that the individual outcomes were independent, and they were reliably documented. These basic issues seem more important than the specific statistical method adopted.

New concerns arise, however, when we go outside the model for the trial data and start to think of applying the analysis to the outside world. Remember that the initial question was about the effect of the dexamethasone on the survival of patients who are severely ill with Covid-19. But careful reading will reveal that we have not quite answered that question. I have been careful to describe the comparison as being in terms of the *randomized groups* in the trial, rather than the effect of the actual treatment when given in normal clinical circumstances – this is a different 'object' of uncertainty, and raises two issues.

First, the trial did not include all eligible patients; dexamethasone was unavailable in 15% of patients, and in 3% the clinical team felt it was either necessary or should not be given, and so did not randomize the patients. Second, and more important, is the fact that the analysis was by 'intention to treat', in that patients were kept in the group to which they had been allocated, regardless of whether they did or did not receive the treatment. So the object of uncertainty in the study is not the effect of *receiving* dexamethasone, but the effect of being *randomized to be given* dexamethasone. It turned out that, of those randomized to receive dexamethasone, 5% of the patients did not actually receive a glucocorticoid, while in the 'usual care' control group, 8% went on to receive a glucocorticoid as part of their clinical care. So there was some 'contamination' between the groups.

The published paper does not report the mortality rates according to the treatment actually received, but if we assume those who did not receive their allocated treatment were essentially chosen at random, we would estimate the effect of giving the treatment to be a little larger than that reported in the paper. This adjustment could be biased if the treatment was influenced by the severity of the illness of the patient, and if we wanted to make claims about the effect when actually used, we should really introduce some additional uncertainty.

It is worth adding that this trial, conducted efficiently and rapidly at the start of the epidemic, led to a major influence on treatment. It was later estimated that in the nine months after the RECOVERY team reported their results, dexamethasone, an inexpensive, readily available steroid, saved an estimated 1 million lives worldwide, including 22,000 in the UK.[12]

————

While randomized trials such as RECOVERY are considered the 'gold standard' for evaluating new treatments, a less-valued option is to simply compare the outcomes of patients who happened to receive a treatment or not. We need to be very careful in interpreting the results of such *observational* studies, and it is helpful to distinguish two broad types of bias: internal and external.

- Internal biases affect the *rigour* of the study, in the sense of its ability to accurately estimate what it is trying to measure. While a randomized study should have minimal internal biases, as the groups are balanced and data is collected according to a rigorous protocol, observational studies will not have a proper control group, and will generally be using routine data-sources.
- External biases affect the *relevance* of the study, in the sense of its generalizability to your question of interest. In the dexamethasone trial they were using 'as-randomized' groups while we are really interested in 'as-treated' comparisons, although this had a minor effect. But in observational studies the population, intervention and the outcome measure may not correspond to the effect you are really interested in.

These limitations in observational studies mean that the uncertainty intervals calculated using standard statistical methods, whether classical or Bayesian, will generally be far too narrow.

One solution is to apply the ideas of metrology, quantifying our 'Type B' subjective uncertainty and adding it to the analysis that is conditional on all the assumptions being correct. I was in a team that examined a series of observational studies evaluating a preventative treatment for pregnant women with a rhesus-negative blood-type, which required us to make judgements about the sizes of potential biases; for example, we assessed that the internal biases in one study meant that the effect may be overestimated by between 20% and 65%.[13]

Such judgements served to expand the width of the intervals and bring apparently conflicting studies into alignment.

This procedure requires people to be upfront about how big they think potential biases could be, based on careful consideration of all aspects of the study. I feel this could be just as valuable as the elicitation of plausible treatment effects described in Chapter 3.

Do we need to choose a model anyway?

The phrase 'model uncertainty' is often used for the common situation when we don't know which model to adopt. But this seems an inappropriate term, since we can almost never conceive of a situation in which a 'true' model is miraculously revealed. So selecting a model (if we do want to do this) is a *decision* which, as we shall see in Chapter 15, means it will be influenced by numerous contextual factors. These might include practical considerations of computing time, its explainability to others, its robustness to unprovable assumptions, and whether it captures the features necessary for the task in hand.

An important lesson is that, to avoid excessive focus on a single story by choosing one single model, we should keep a diversity of perspectives in mind, learning from their agreements and differences. Perhaps the ideal, although resource-intensive, solution is to have multiple independent teams developing their own models to tackle the same problem, just like Obama had multiple teams assessing the probability that bin Laden was in the Abbottabad compound. And this is exactly what happened in the UK during the Covid-19 pandemic.

In the Covid-19 pandemic, what was the median value for *R* in the UK on 14 October 2020?

During the Covid-19 pandemic, we repeatedly heard about the current estimated value for *R*, which is the average number of people that someone with the virus goes on to infect. This is a standard metric for monitoring the progress of an epidemic, since if $R > 1$ the epidemic is growing, and if $R < 1$ it is contracting. It is impossible to directly observe values for *R* and so it needs to be estimated through complex statistical modelling, and throughout the UK multiple pandemic teams provided estimates using a wide range of approaches and sources of data, from mathematical models for hospital admissions to 'agent-based' models that simulate what is going on for all individuals in a population.[14]

R varied considerably around the country, and the median value across the UK is the headline figure that received most attention. Figure 8.3 shows 90% confidence intervals for the estimates of the median *R* for twelve different models, as featured in a 'consensus statement' by the Scientific Pandemic Influenza Group on Modelling, Operational subgroup (SPI-M-O) on 14 October 2020.[15]

The estimates exhibit substantial disagreement, with many of the intervals not even overlapping – since they are all trying to estimate the same quantity, this immediately demonstrates that at least some of the published intervals are far too narrow. But it is hardly surprising that the intervals are overconfident, since they are calculated on the assumption that each model is 'true', which of course we know is not the case. It's important to remember that simpler models, with more assumptions and fewer parameters to estimate, tend to give tighter intervals, and this can give a misleading impression of confidence. So narrow intervals do not generally represent a 'good' model, but just a simpler one with possibly more biases.

The SPI-M-O group were then faced with the task of combining all these diverse results into a single consensus opinion. One approach might have

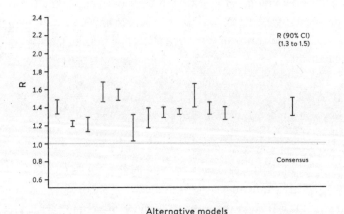

Figure 8.3
SPI-M-O estimates of median *R* in the UK, expressed as 90% confidence intervals. The final interval is the combined range after rounding to one decimal place. Note the substantial variation between the intervals, many of which do not overlap.

been to consider the different models as if they were 'experts' each giving their opinion about R, and to construct a (fairly dispersed) composite opinion by averaging the distributions represented by the intervals. An alternative, chosen by the SPI-M-O group, is to essentially consider the results as independent studies contributing their own 'data' which is then combined using a standard technique known as a random-effects meta-analysis. This is usually used to combine data from multiple clinical trials trying to estimate the effectiveness of the same drug, where allowance is made for the treatment effects to differ between studies – although these do not generally show the massive variation of the R estimates. The final conclusion is shown in the final interval in Figure 8.3, and the consensus estimate and interval are claimed to be fairly robust to the precise method used.[16]

This example demonstrates the value of including multiple perspectives – taking a single model at face value could have grossly overstated our confidence, since it is conditional on one particular set of assumptions. Multiple analyses demonstrate the sensitivity of the results, an extreme example being when two sets of ecology data were analysed by 246 different biologists, who reached remarkably different conclusions about blue-tit nesting behaviour and eucalyptus seedlings, even after poor analyses were excluded.[17] When results from a wide range of independent teams are pooled, the resulting uncertainty may arise as much from disagreement between different groups as within-model uncertainty.

We shall return to this issue when we face the highly contested area of climate modelling in Chapter 10. But even in the biggest of scientific endeavours, there is still room for appropriate caution in claiming a 'discovery'.

How sure can we be that the Higgs boson exists?

The 'Standard Model' is the current best theory for the fundamental structure of matter and forces in the universe, but for decades physicists faced the problem that a basic component – the Higgs boson – had not been shown to actually exist. Eventually, complex and very expensive experiments were carried out at the Large Hadron Collider at the European Centre for Nuclear Research (CERN), in which events (specific particle collisions) were counted for different masses of particles; if the Higgs boson did not exist, it was assumed these would follow a Poisson distribution around a smooth background line, while if the Higgs boson did exist and had mass m_H, an excess number of events would be expected around m_H. In 2012 two separate research groups reported their results, and their graphs clearly displayed a bump around a mass of $126 \, GeV/c^2$, just where it would have been expected from the theory.

Such an observation may look convincing to a casual reader but is insufficient to declare a fundamental discovery in the laws of physics, which depends on formal statistical analysis. A 'local' P-value was calculated for each potential mass, representing the probability of getting such an extreme count under the null hypothesis that the Higgs boson did not exist.[18]* Particle physicists tend to measure the incompatibility of their findings with a null hypothesis in terms of 'sigmas'; for example, a '2-sigma' result is equivalent to observing a statistic that is 2 standard deviations greater than its expectation under the null hypothesis, which, assuming a normal distribution, corresponds to a P-value of 0.025, which in many contexts might be considered fairly strong evidence. However, particle physicists make far more stringent demands, requiring at least a 5-sigma result, corresponding to a P-value of 1 in 3.5 million. Fortunately the two independent teams found 5- and 6-sigma results for their local P-values,† and further work led to an announcement by CERN in 2013 that the evidence about the particle 'strongly indicates that it is a Higgs boson'.[19]

So why does the physics community require such strong evidence? First, they are very keen to avoid the embarrassment of making a 'false discovery' – a public claim that has to be later retracted. Second, as we have repeatedly emphasized, any P-value is calculated assuming both the null hypothesis and all the other assumptions in the model are true, and the models in the Higgs investigation contain numerous details and approximations that are acknowledged not to represent reality. Third, the final P-value is the smallest of all the local P-values at a range of masses, and allowance should really be made for this multiple testing, known in physics as the 'look elsewhere effect'.

The quoted P-value is therefore not claimed to be a precise probability, but rather a broad measure of the compatibility (or lack of it) of the data with a null hypothesis. The choice of 5-sigma is therefore a rather ad hoc threshold

---

* This was based on a likelihood-ratio test assuming a Poisson distribution. Simultaneously a P-value was calculated under the null hypothesis that the Higgs did exist, hoping for a P-value near 1, suggesting compatibility of the data with the theory. This was intended to ensure that the data did not just show incompatibility with the non-existence of the Higgs (using the convoluted logic of P-values), but actively supported its existence.

† Predictably, the P-values were often incorrectly reported as the probability of the null hypothesis being true, for example in Forbes magazine: 'The chances are less than 1 in a million that it's not the Higgs boson.'

for claiming a discovery, rather than a formal expression of uncertainty. And even a 5-sigma result will need further replication and confirmation before becoming accepted; for example, in 2003 the so-called pentaquark particle was discovered at 5.2 sigma[20] but later became completely discredited,[21] while the 6-sigma finding of faster-than-light neutrinos in 2011[22] had to be retracted the following year when the results were found to be due to equipment failures.

So how sure can we be about the Higgs boson? We could assess likelihood ratios (known as **Bayes factors** in this context), comparing the evidence for and against the theory. It would in principle even be possible to produce a (subjective) probability that it exists, but this would require strong assumptions about prior probabilities before the experiments were carried out. But presumably the scientific community was confident enough to finally jointly award Peter Higgs a Nobel Prize in 2013, fifty years after he and others proposed the particle.

––––––––

All of the ideas in this chapter stay within the basic paradigm of statistical inference, in which assumed probabilistic models for what we observe lead to expressions of uncertainty about our conclusions. Vast numbers of scientific studies report their results in terms of confidence intervals and P-values, usually based on statistical packages for standard methods such as regression analysis. More bespoke modelling may simulate Bayesian posterior distributions which are summarized as estimates and so-called 'credible intervals'.

A whole research area of *uncertainty quantification* (UQ) has developed, concerned with ways of constructing probability distributions for unknowns, measuring sensitivity to important sources of uncertainty, and identifying how our uncertainty might change were we to gain some piece of additional evidence. This work can become very technical, and computationally impractical for some extremely large models of, say, oil-field reserves. So *emulators* may be built, allowing rapid estimation of what the model would have come up with had we sufficient time and resources. These are essentially models of models of the real world.

Concern is sometimes expressed about Bayesian methods introducing subjectivity into science, and there have been repeated efforts to develop 'objective' Bayes methods. But, as statistician Andrew Gelman has pointed out, the choice of analysis is itself a personal judgement.[23] Instead of trying to divide approaches to statistical modelling into 'subjective' and 'objective', we should instead emphasize both 'objective' features such as transparency, impartiality and how well the models represent external reality, as well as more 'subjective' characteristics such as the role of judgement and acknowledgement of multiple perspectives.

So experienced researchers need to have the humility to acknowledge that any statistical model is not a completely accurate description of reality, and that therefore any resulting uncertainty assessments are never 'correct'. We have seen five main approaches that people take to deal with this challenge:

1. Express any *caveats* clearly and prominently.
2. Do *sensitivity analysis* to different choices of models.
3. *Combine the results of a wide range of models*, preferably from independent teams, to try to avoid depending on a single perspective.
4. As recommended in metrology, *elaborate the model to include 'Type B' subjective probability assessments* so that, for example, potential biases in the data are taken into account.
5. Do the standard calculations of P-values, and so on, but take them more as *indicators* rather than accurate probabilities.

Personally, I feel that we should do our very best to model the world, and then it is fine to introduce Type B subjective judgement about the limitations of our model. But in the end, models are fictions, just metaphors for reality, and sometimes we should just admit we don't fully understand what is going on. And in the next chapter we explore attempts to be upfront about such lack of confidence in our understanding.

Summary

- Epistemic uncertainty about states of the world is based on evidence from data. We need assumptions about how what we observe relates to true underlying states, and these form the basis for a statistical model.
- Statistical methods turn assumptions about variability into statements of uncertainty about aspects of the model which correspond to states of the world. We can quantify our epistemic uncertainty as intervals or distributions, depending on whether we take a 'classical' or Bayesian perspective.
- However, this assessment of uncertainty is conditional on the truth of the model, which we know is not the case.
- Once we acknowledge that calculations of uncertainty based on a single model may be optimistic, we can check sensitivity to different models, combine results from multiple teams, use judgement to elaborate the model to allow for potential inadequacies and biases, or acknowledge that our measures are only indicators.
- Even then, we may feel we need to express caveats about our numerical assessment of uncertainty.

CHAPTER 9

How Much Confidence
Do We Have in Our Analysis?

I let you examine a coin, it looks fair, and then you flip it repeatedly and it comes down heads around half the time. If I ask you for your probability that it will come up heads when next flipped on to a hard surface, my guess is that you will say '50%'. But suppose I show you two apparently identical coins, A and B, and ask you for your probability that coin A is heavier, even if only by a tiny fraction of a gram. You presumably will have no idea, although if pushed you may reluctantly say '50%', just because you have no reason to choose between them. These two assessments are numerically identical but qualitatively quite different – the first is based on informed judgement, while the second is completely lacking any evidence. You will presumably feel more confident in the first.

Intelligence analysts face similar challenges, and in a somewhat more important context. We've already seen in Chapter 2 how various agencies encourage the assessment of numerical probabilities, which might then be communicated using a scale such as the UK Probability Yardstick, where, for example, probabilities between 55% and 75% correspond to the verbal term 'likely'. But what if such an assessment is based on mere fragments of poor-quality evidence, and the analyst knows they are missing some vital, and potentially obtainable, information? The UK Ministry of Defence[1] recognizes that analysts will feel a lot happier with some assessments than others, and recommends they explicitly judge their 'analytic **confidence**' about the robustness of any probability assessment. This will depend on the quality and quantity of the evidence available, the rigour of the analytic process, and the complexity and volatility of the situation.

The US National Intelligence Council make very similar recommendations, saying that 'Intelligence Community judgments often include two important elements: judgments of how likely it is that something has happened or will happen . . . and confidence levels in those judgments (low, moderate, and high) that refer to the evidentiary basis, logic and reasoning, and precedents that underpin the judgments.'[2] Intelligence reports depend crucially on the trustworthiness of sources, and so it is unsurprising that US intelligence analysts are able to express *low confidence*, meaning 'that the information's credibility and/or plausibility is uncertain, that the information is too fragmented or poorly corroborated to make solid analytic inferences, or that reliability of the sources is questionable'.

Intelligence analysts are understandably unwilling to give confident numerical judgements if they feel that their estimates might change substantially when more evidence arrives in the future – so-called 'information gaps'.[3] They are not alone; doctors avoid making prognoses until they have done important tests and, more mundanely, you may hesitate to estimate how long a train journey may take until you know whether any strikes or engineering works are planned.

But, as we shall see in other domains, the terms are not used consistently.[4] Although 'confidence' is supposed to *supplement* a numerical measure of probability, it often *substitutes* for it. For example, in 2017 all three US intelligence agencies agreed that Putin and the Russian government aspired to help President-elect Trump's election chances by discrediting his opponent, Hillary Clinton; the CIA and FBI had *high confidence* in this judgement, while the National Security Agency had *moderate confidence*.[5] Perhaps the imprecise nature of such a claim discourages analysts from using a 'likelihood' scale, in contrast to more precise objects of uncertainty such as Osama bin Laden's presence in the Abbottabad compound, discussed in Chapter 2.

The intelligence community is not alone in valuing judgements about analytic confidence. We shall see that many different groups of researchers have developed their own scales, applying them to whole statistical analyses, reflecting the challenges of answering important questions with only limited pieces of evidence. For example, while I've been in teams working on some fairly difficult problems, rife with uncertainty, I think the most challenging question that I've been asked is:

In the UK between 1970 and 1991, how many people were infected with hepatitis C through transfusions of contaminated blood?

In the 1970s and 1980s, many people received transfusions of contaminated blood and then contracted diseases such as HIV/AIDS or hepatitis C. In particular, people with haemophilia were given blood products which had been concentrated from many donations, some from US prisoners who had been paid to give blood. If only one donor in a pooled sample had HIV, then the entire batch would be contaminated. The resulting international scandals led, for example, to the head of the French National Centre for Blood Transfusion being sentenced to four years in prison in 1992. In the UK, this was referred to in the House of Commons in 2017 as 'the worst treatment disaster in the history of our NHS, and one of the worst peacetime disasters ever to take place in this country'.[6]

After years of campaigning by those affected, the Infected Blood Inquiry was set up in 2018, and I was part of the Statistics Expert Group charged with estimating both the number of people infected, and the number of people who subsequently died because of their infection. These are historical events, happening up to forty years ago, and so the uncertainty is purely epistemic.

Some conclusions could be reached with considerable confidence. For example, existing databases and registries of claimants for compensation broadly agreed that around 1,250 people with bleeding disorders such as haemophilia were diagnosed with HIV from 1979 onwards, with a peak in 1985. Around three quarters had died by 2019, around half due to HIV-related causes. This was a massive tragedy.

It is far more difficult to estimate the number receiving ordinary blood transfusions who were infected with hepatitis C (HCV) before testing for HCV became available in 1991. Although chronic HCV infection can lead to liver cancer, liver failure and other serious conditions, a long incubation period means that many individuals who received contaminated blood may never have known they were infected and would not feature on any registry, since any HCV diagnoses are likely to be many years after the HCV-implicated transfusion.

So instead of counting specific (although anonymous) individuals, we had to use a complex statistical model for the whole process, going from estimates of the proportion of infected donations and the number of infected transfusions, through to the numbers being chronically infected, and the long-term consequences of infection. Using the language of metrology, we needed to take account of both Type A (statistical) and Type B (judgemental) uncertainty.

For example, an important input to the modelling was an estimate of the percentage of people infected with HCV who naturally clear the virus and do not go on to become chronically infected. There was good published data on this,[7] allowing us to represent our uncertainty by a normal distribution with mean 18% and standard deviation 3%. But there were no relevant data for some other parts of the model, and so expert judgements had to be used.

These numerous sources of uncertainty were taken into account when coming up with a final estimate of the number of infections and deaths. Each unknown parameter was assigned a probability distribution to produce a 'stochastic' model, and then the model was run 10,000 times – for each run, values for each parameter were simulated from its specified distribution and then propagated through the model. This produced 10,000 plausible values for each outcome, which were summarized by their median and 95% uncertainty intervals, as shown in Table 9.1.* This is a standard Monte Carlo approach, as introduced in Chapter 6, and sometimes known as a *probabilistic sensitivity analysis*. Note that we use the term **uncertainty interval** to distinguish it from a confidence interval calculated as part of a standard data analysis.

Table 9.1 shows that the model estimated that around 27,000 people had been infected in the UK, but with considerable uncertainty. The number of attributable deaths was high, at around 1,800, but again with very large uncertainty. It is important to note that we did not have any records of who these individuals might be.

The large number of uncheckable assumptions meant the Statistical Expert Group wanted to express considerable caution about our whole analysis, in particular the estimates and intervals in Table 9.1. So we adopted the scale used in scientific advice during the Covid-19 pandemic (see below), and said that we only had *moderate confidence* that the available data could answer the questions we had been asked. It was liberating to be able to use this scale; for example, when asked for the number of people who had been infected with hepatitis B, we found there was little data and no reliable model we could use, so we said we had *low confidence* in being able to answer the question, and refused to provide any numbers.

Our final evidence to the Infected Blood Inquiry estimated that around 3,000 people had died due to receiving infected blood or blood products, including many young people. We made no attempt to evaluate the huge harm done to the families of the victims, some of whom sat patiently in the front

---

* The model was implemented by Ruth McCabe and Sarah Hayes in the statistical programming language R.

Quantity of interest	Median estimate	95% uncertainty interval
Number of people in the UK infected with HCV through blood transfusion between January 1970 and August 1991	26,800	21,300–38,800
Number of people chronically HCV-infected and who had died of any cause by the end of 2019	19,300	15,100–28,200
Number of deaths by the end of 2019 attributable to HCV infection	1,820	650–3,320

Table 9.1
Median estimates and 95% uncertainty intervals of major quantities of interest for the UK from the statistical model of HCV infections from blood transfusions.[8]

row while we spent a day answering detailed technical questions from the inquiry team. While we admitted substantial uncertainty about the precise numbers, we could be confident that a great harm had been done.

Direct and indirect uncertainty

I chose the infected blood example to illustrate a range of challenges: the complexity of a reasonably realistic model, the use of Monte Carlo simulations, the value of both statistical (Type A) and judgemental (Type B) quantification, and different types of sensitivity analysis. And we were only one team with one model – who knows what variation would have been seen if completely independent groups had tackled the problem? But the primary aim of this example was to illustrate the use of a qualitative 'confidence' scale to express the remaining doubt that we had about the quality of the evidence, the appropriateness of our model and the accuracy of our results.

Suppose a scientific claim is being made, which may be a fact, an estimate, a trend, and so on. We've seen many examples of using statistical models to assess what we call **direct uncertainty** about the claim – this may take the form of a probability, an interval, or a distribution. But the infected-blood example showed that, even after exhausting our efforts at quantification, we can still be left with doubts about our analysis. This calls for an additional way of expressing **indirect uncertainty**, relating to the strength and quality of the available evidence.

As I pointed out earlier, researchers in many areas have independently found a need for similar measures, although their use has not always been consistent or clear. For example, we saw in Chapter 2 how the Intergovernmental Panel on Climate Change (IPCC) uses a scale of 'likelihood' to translate numerical probability assessments into words, and vice versa; for example, the term 'likely' means between 66% and 100%. But as well as these measures of direct uncertainty, the IPCC also recommends[9] using a level of 'confidence', on a scale *very low, low, medium, high, very high*, which summarizes the judgements of the team about the validity of a claim, in terms of the strength of the evidence and expert agreement.

For example, the IPCC Sixth Assessment Report's 2021 summary for policymakers on the physical science basis of climate change[10] includes claims:

- with direct uncertainty alone; such as the 'likely range of total human-caused global surface temperature increase from 1850–1900 to 2010–2019 is 0.8°C to 1.3°C, with a best estimate of 1.07°C'. The IPCC says there is no need to mention high or very-high confidence in probabilistic statements, so presumably the high level of confidence is implicit.
- with both direct and indirect uncertainty, such as 'Globally averaged precipitation over land has likely increased since 1950, with a faster rate of increase since the 1980s (medium confidence).'
- with just indirect uncertainty, such as 'In 2011–2020, annual average Arctic sea ice area reached its lowest level since at least 1850 (high confidence)', and 'There is low confidence in the projected decrease of Antarctic sea ice.'

In the second example above, the confidence measure is used as a supplement to the probability assessment, while in the third bullet point, confidence is used as a substitute for the direct 'likelihood' terms, when the authors presumably did not feel able to assess a probability. This has led to concern about whether these terms are used clearly and consistently throughout the IPCC's publications.[11]

The medical world has also seen the need for the 'GRADE' scale, of *very low, low, moderate, high* quality evidence. For example, a 2010 review estimated that radiotherapy after surgery for cervical cancer reduced the risk of disease progression by 42% (95% confidence interval 9% to 63%), primarily based on a single well-conducted randomized trial with only around 280 participants – the GRADE assessment of the quality of this evidence was *moderate*. There is a formal process for deciding the appropriate GRADE level, taking into account the risk of bias, imprecision, inconsistency, indirectness and publication bias, although assigning a level still requires considerable judgement. GRADE is used by over a hundred organizations worldwide.[12, 13]

The rating scale used to be defined in terms of whether further research was likely to change the result, but in 2011 the guidance changed[14] and GRADE is now referred to as a *certainty of evidence* scale, with the levels defined as

- Very low: *The true effect is probably markedly different from the estimated effect.*
- Low: *The true effect might be markedly different from the estimated effect.*
- Moderate: *The authors believe that the true effect is probably close to the estimated effect.*
- High: *The authors have a lot of confidence that the true effect is similar to the estimated effect.*

So although GRADE is constructed as a quality-of-evidence scale, its interpretation is now as direct uncertainty – substituting for the model-based uncertainty around the estimated effect, rather than supplementing it. This is rather like the effect of adding Type B subjective uncertainty, but without quantifying the effect.

Another contested area of science arose in the Covid-19 pandemic, where the potential benefits and harms of lockdowns, masks, vaccines, and so on were vigorously debated, and continue to be so. The UK Scientific Advisory Group for Emergencies (SAGE) met frequently throughout the pandemic (I attended one of their meetings), and their judgements about the effectiveness of different policies to reduce the spread of the virus were generally accompanied by a summary of confidence, on a scale of *low, low–moderate, moderate, moderate–high, high* – the same scale we adopted for our infected blood analysis. For example, in September 2020 they concluded that restrictions on outdoor gatherings, including prohibiting large events, would have a low impact on Covid transmission with *high confidence*, whereas closing all schools would be associated with a reduction in R of 0.2–0.5, but with *low confidence.*[15]

Unfortunately, even when scientists acknowledge doubts about their own knowledge, this is often not widely communicated or understood – politicians who decide policies tend to act as if the evidence behind their decisions is irrefutable. To counter this, some organizations are making a prominent feature of their uncertainty. For example, the UK Education Endowment Foundation provides guidance about policies for improving education, and their Teaching and Learning Toolkit[16] displays their judgements rather like a hotel star-rating scheme, with one to five little 'padlocks' representing their confidence in the conclusions. And the UK Statistics Authority uses the label 'Official Statistics in Development' for those that cannot yet be considered as 'official statistics', and which may be subject to 'a wide degree of uncertainty in new estimates or increasing the uncertainty in existing statistics'.[17]

The lessons learned from all these examples are summarized in Figure 9.1, showing both direct and indirect pathways to expressing uncertainty about a claim of a fact, estimate, trend, or causal connection.

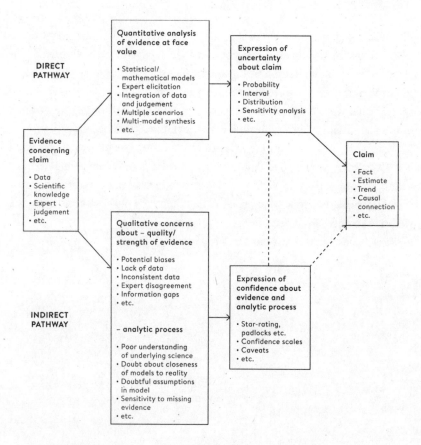

Figure 9.1
Direct and indirect pathways to uncertainty about a claim of a fact, an estimate, a trend or a causal connection. Direct uncertainty arises from statistical modelling or expert judgement. Indirect uncertainty arises through concerns about the quality and strength of evidence, and the whole analytic process, and a summary may be applied to either the quantitative expression of uncertainty, or to the claim itself (dashed arrows).

There could be many concerns about the basis for the claim – the data may be of poor quality and have potential biases, experts may not agree and, crucially, there may be important information gaps. There is often an unfortunate lack of clarity about how these scales should be used. Although the object of the 'confidence' is generally described as the whole analytic process, we've seen that the terms are often applied to the claim itself, apparently as a non-numerical degree of certainty, being used as a substitute for a probability, rather than a statement about the assessed probability.

––––––––

The scales we have looked at in this chapter are popular because analysts are often, very sensibly, unwilling to commit themselves fully to numerical conclusions based on models alone. Our examples have all concerned our lack of scientific knowledge – pure epistemic uncertainty – where we can generally specify what we don't know. In Donald Rumsfeld's eternal phrase mentioned in the Introduction, we are dealing with the 'known unknowns'.

Yet sometimes we cannot conceptualize all the possibilities. As we shall see in Chapter 13, such 'deep uncertainty' may be faced when trying to make long-term predictions but can also occur with lack of knowledge – the epistemic uncertainty about what type of alien lifeforms may exist has not got a well-defined list of options.

Sometimes we may just have to admit we don't know.

Summary

- Even after making every effort with quantitative modelling and judgement, many researchers and organizations feel the need for additional scales of 'confidence'.
- These scales arise from an 'indirect' pathway to uncertainty about a claim, which may reflect the quality of available evidence, the degree of expert agreement and acknowledged gaps in information.
- Qualitative 'confidence' scales are used both as a supplement to a numerical assessment of uncertainty and as a substitute when people are unwilling to quantify their uncertainty.
- In spite of the widespread use of such scales, there is often lack of clarity about their precise meaning.
- Even these scales will be inadequate when dealing with situations when we cannot even list the possibilities.

CHAPTER 10

What, or Who, is to Blame?
Causality, Climate and Crime

You walk into a room, flip the only light-switch, and the light comes on. This is the simplest type of causality – the basic physical mechanism determines the light comes on if, and only if, you flip the switch. Things can, of course, rapidly get more complicated – there may be more than one switch, the circuit may be faulty, the bulb may have gone – but it should be possible to assess what is going on using observation and logic.

In this chapter we will look at two rather more challenging situations. The first concerns *general causation* – whether some action or exposure A tends to cause the outcome B in repeated situations, in the sense of raising the assessed probability of B occurring. Typical examples concern whether particular foodstuffs raise the risk of cancer, or vaccines are a major cause of harm. Essentially this examines uncertainty about 'the effects of causes', essentially answering 'what-if?' questions.

We then look at *specific causation*, which considers an individual event and asks whether, or to what extent, a previous action or exposure led to that event occurring. This is now looking at 'causes of effects', also known as *attribution*, essentially answering 'why?' questions. We humans are quick to explain why something happened, whether it's a road accident, a heart attack, the break-up of a relationship, or the unexpected result of a referendum. Everyone can loudly proclaim their own theory, in spite of Michael Blastland's 'hidden half' meaning that there is generally no simple explanation for events. There's a popular fallacy known in Latin as *post hoc ergo propter hoc*: after observing that B followed A, conclude that A actually caused B to happen. A classic example is when a football team has suffered a string of defeats, they sack the manager, win the next match, and then people claim the sacking changed the team's fortune. But we have already seen there is a huge amount of luck in football, and so this may be a period of bad luck coming to an end.*

In keeping with the rest of the book, we shall avoid such personal intuitions and instead focus on questions where an analytic approach to attribution is reasonable, such as legal cases in which people claim to have been harmed by exposure to particular chemicals, recent arguments about whether man-made climate change was responsible for an episode of extreme weather, or even court investigations of whether an individual was responsible for a crime.

Claims about both general and specific causation are dominated by uncertainty. This may be just a verbal qualification, as in 'violent video games may increase the risk of aggression',[1] but here we look at attempts to be more rigorous in expressing uncertainty, either numerically or at least on a formal scale. The discussion will inevitably get a bit technical, although hopefully these ideas can help resolve competing claims on some deeply contested issues.

General causation

We've already seen in Chapter 8 how randomization allows us to assess causality – by comparing the outcomes in groups of people who have been randomly assigned to receive dexamethasone or a control, we can be sure, up to the play of chance, that any differences will be due to the assigned treatment. We have not just observed that the probability of a good recovery is higher in a group receiving the drug, we have actively intervened to create groups that will be matched, even for factors that we are unaware may influence outcomes.

But the media love clickbait headlines like 'Can the cat give you cancer?',[2] and these are certainly not based on randomized trials. So how confident can we be about causation, when there has been no experiment? The next example shows that great care is needed when making such claims.

---

* This is known as *regression-to-the-mean*, and I watched it happen with my local team Cambridge United in December 2023.

Does hormone replacement therapy (HRT) harm women?

Hormone replacement therapy (HRT) is generally given to relieve the serious symptoms experienced by women going through their menopause, but many observational studies, in which large numbers of people are followed up over time, have shown that hormone replacement therapy (HRT) is also associated with better cardiovascular outcomes. These are only 'correlations', but causal language has often been used, for example a much-cited 1992 review stated that there is extensive and consistent evidence that HRT use 'reduces risk' for coronary heart disease by about 35%.[3]

But was this really causal, or did women who take HRT, who tend to be younger and at or soon after the menopause, tend to be at lower risk anyway? When the results of a large Women's Health Initiative randomized trial were published in 2002, they showed that HRT *increased* the annual risk of coronary heart disease by 18% (95% interval: 5% decrease to 45% increase), as well as increasing the risk of invasive breast cancer, stroke and pulmonary embolism.[4] This caused widespread consternation, to put it mildly, and a substantial fall in prescribing HRT.

But further analysis has largely resolved the apparent contradiction, as the observational and randomized studies had looked at different groups. Researchers are now confident that, if initiated in most women who are under sixty years of age or at or near menopause, a limited period of HRT significantly reduces mortality and cardiovascular disease,[5] and for these women the benefits may outweigh the risks. It's all about the timing.

———

The HRT example shows the care and complexity needed in assessing causality, particularly in the absence of randomized trials. A classic example is the many years it took for the association between smoking and lung cancer to finally be determined to be causal, based on numerous lines of research, and despite the efforts of the tobacco industry to cast doubts on the science.

Once we can assume a causal relationship, we can try to answer questions such as what proportion (and with what uncertainty) of lung cancers could have been avoided had people not smoked? For example, in a study of Norwegian women,[6] current smokers had a relative risk for lung cancer that was fourteen times that of never-smokers (95% interval 10 to 19). This means that of every fourteen smokers who get lung cancer, one would have got it anyway, and so thirteen (93% of the total) got it because of smoking. This is known as the **attributable fraction** or **excess fraction**, and from this study will have a 95% interval of 90% to 95%. In notation, if RR is the relative risk, the attributable or excess fraction is $AF = (RR - 1) / R = 1 - 1 / RR$, which in this study was $1 - 1 / 14 = 0.93$.

But we wanted an idea of the proportion of all cases of lung cancer caused by smoking, which is known as the **population attributable fraction**. For this we need to know the proportion of women who ever smoked, which over the period of this study was around 30%. From this we can estimate that the population attributable fraction is 80% (with 95% interval of 73% to 84%).* This means that, in principle, around 80% of female cases of lung cancer (as well as many other life-threatening conditions) could have been avoided had people not smoked.

Measures like these demonstrate the potential benefits of changing behaviour, and so reducing the impact of 'effects of causes'. But, as we will see later, the attributable fraction can also provide a basis for attributing 'causes to effects', and so be used in legal rulings on claims for compensation.

———————

While the causal connection between smoking and lung cancer has been established beyond reasonable doubt, other causes of cancer are not so clear. The International Agency for Research in Cancer (IARC) has a long-standing programme of investigating whether large numbers of chemicals and other exposures are carcinogenic (meaning capable of causing cancer in humans), and after extensive research classify each into one of four categories:

- *Group 1: carcinogenic to humans.* Examples include plutonium, ionizing radiation, working as a firefighter, smoking, alcohol, processed meat.
- *Group 2A: probably carcinogenic to humans.* E.g. working as a hairdresser or barber (due to exposure to certain chemicals), working on night shifts, very hot beverages, red meat.
- *Group 2B: possibly carcinogenic to humans.* E.g. aloe vera, working in dry-cleaning.
- *Group 3: not classifiable as to its carcinogenicity to humans.* E.g. coal dust, coffee, silicone breast implants.

———————————————————————————

* If P is the prevalence of the causal risk factor, then the population attributable fraction is $(PAF) = P(RR - 1)/(P(RR - 1) + 1)$.

There's been a lot of misunderstanding about these classifications. Processed meat and smoking are both in Group 1, but this does not mean they are equally dangerous, despite grossly misleading headlines such as 'Bacon, ham and sausages have the same cancer risk as cigarettes, warn experts'.[7] This is because the IARC classification concerns *hazard* and not *risk*, where, as we saw in Chapter 1, hazard is the potential, in possibly very extreme circumstances, to cause harm, whereas in this context risk means the actual chance of harm, given the normal way we live. So when the IARC put processed meat in Group 1 as 'carcinogenic to humans', along with smoking, ionizing radiation and plutonium, they certainly did not imply they were the same risk.

The IARC has tried to improve its explanation of what its classification means, but this has not stopped its judgements from being misunderstood. For example . . .

> **Can aspartame, an ingredient of drinks like Diet Coke, give you cancer?**

Aspartame is a low-calorie artificial sweetener that has been used for decades in a huge range of foodstuffs, notably diet drinks. In 2023, based on a wide range of evidence, the IARC put aspartame into Group 2B, *possibly carcinogenic to humans*, which, according to their published algorithm,[8] means that at least one of the following criteria has been established:

- *Limited evidence of carcinogenicity in humans*
- *Sufficient evidence of carcinogenicity in experimental animals*
- *Strong evidence that the agent exhibits key characteristics of carcinogens*

Unfortunately, there was a leak to the media two weeks before the official announcement by the IARC, which resulted in headlines like 'Aspartame sweetener to be declared possible cancer risk',[9] making exactly the error that had been made about bacon. This was particularly ironic, since the announcement of the IARC classification was made simultaneously with a statement from the WHO Food and Agriculture Organization Joint Expert Committee on Food Additives (JECFA) about the actual risk, which found 'no convincing evidence from experimental animal or human data that aspartame has adverse effects'.

These two announcements may appear to be contradictory, but it is quite feasible that aspartame could be carcinogenic if taken in sufficient quantities and yet is of no measurable risk at the amount humans consume. Advice by the WHO has remained unchanged for forty years: it still says on average people are safe to drink up to fourteen cans of diet drink a day (about half a bucketful). So, to answer the question posed at the start of this section, you are not going to get cancer from diet drinks – although doubtless there will be court cases about this.

The IARC classification uses terms like 'probably', but confusingly this is not a direct expression of probability of carcinogenicity. It is a qualitative assessment of the strength of evidence for carcinogenicity, and so is more an expression of 'indirect uncertainty', as discussed in Chapter 9. The US Environmental Protection Agency (EPA) reinforces this approach,[10] saying, 'most causal inferences are based on the strength of evidence, so that no single source of uncertainty characterizes the uncertainty concerning the conclusion. Therefore, the uncertainty concerning most causal analyses must be characterized qualitatively.'

In contrast, we shall now see that climate change researchers are somewhat bolder, and are willing to put probabilities on their causal claims.

Climate change can be an even more contested area of science and causality than the causes of cancer. There have been years of increasingly polarized conflict between scientists who claim that recent changes in climate are primarily due to human activities, and those who argue that much is simply natural variation. The Intergovernmental Panel on Climate Change (IPCC) has therefore developed a way of expressing its degree of certainty about its causal claims, and their 2021 report[11] uses phrases such as

- 'It is unequivocal that human influence has warmed the atmosphere, ocean and land.'
- 'Human influence is very likely the main driver of the global retreat of glaciers since the 1990s and the decrease in Arctic sea ice area between 1979–1988 and 2010–2019.'
- 'It is likely that human influence contributed to the pattern of observed precipitation changes since the mid-20th century.'

As we've seen in Chapter 2, these 'direct' expressions of uncertainty can be translated to numbers: 'very likely' as 90–100% probability, and 'likely' as greater than 66%. 'Unequivocal' is off the scale, and presumably is to be interpreted as 'certain'.

These are expert judgements based on multiple lines of evidence. A major contribution is based on comparing two mathematical models of the climate: one that projects what we would have expected to happen from 1850 to the present day including the impact of human influences, and another projection of what we might have expected to happen if we hadn't developed our industry and only natural processes were operating. This is known as a *counterfactual*, as it is explicitly concerned with assessing the potential impact of history not developing as it really did.

Figure 10.1 shows such a comparison for global surface annual temperature between 1850 and 2020. Even allowing for the uncertainty in the model projections, the first, human-influenced, model agrees well with what has been observed (black line), while the second 'natural' projection is nowhere near the actual data. These simulated projections are known as 'fingerprints'.

The IPCC also includes a more formal regression analysis to see how well the two models together fit the data – when the coefficient for the

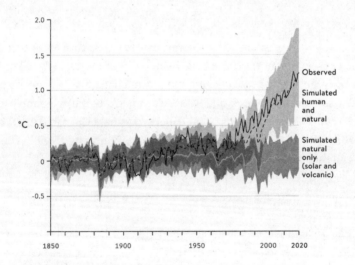

Figure 10.1

The observed change in annual average global surface temperature 1850–2020 (black line), compared to model-based simulations using human and natural (dashed line showing the estimate, and light grey band the uncertainty) and only natural factors (pale line and dark grey band).[12] The simulation including human influences matches the observed data.

'human-influenced' fingerprint is near 1, and the coefficient for the 'natural' fingerprint is near 0, then it can conclude that the man-made warming is approximately equal to the observed warming. The climate researchers therefore carry out a classical statistical analysis, with results reported as estimates and confidence intervals, and use this as a primary basis for a consensus probability judgement about underlying causality, which is essentially a Bayesian idea. And their conclusion was 'unequivocal'.

———————

This analysis considered the possible causes of changes in the *climate* of the entire planet, but it is natural to ask whether specific *weather* events such as heavy rainfall, droughts, heatwaves, and so on can be attributed to man-made climate change. This is also likely to become increasingly important in legal cases about liability for damage from extreme weather.

The following, rather parochial, question is of some personal interest.

> What was the impact of man-made climate change on the record-high temperatures in the UK in September 2023?

I have to admit that I found September 2023 rather splendid, spending many warm evenings outside and taking long cycle rides in the countryside. But while I was smugly enjoying myself, perhaps I should have been wondering about why the almost unprecedented average UK September temperature of 15.2°C had occurred.

The UK Meteorological Office (Met Office) now conducts rapid attribution studies for events such as the record temperatures in September 2023.[13] These are similar to the climate causality assessment described above: a model for the natural variability (NAT) provides a probability distribution for what might be expected in September had there been no human influence, and simulations from models that allow for human influence (HUM) provide a comparable probability distribution. This gives an assessment of the probability of observing such an extreme temperature without and with human influence, which are essentially P-values under the 'competing' hypotheses, denoted P_{NAT} and P_{HUM}.

Figure 10.2 shows the probability distributions under natural and human influences, each generated from multiple runs of the models, and then smoothed. The tail-area P_{HUM} for the HUM distribution was estimated to be 2.7% (90% uncertainty interval 2.4% to 3.1%), meaning the Met Office assesses there was only around a 1 in 40 probability of observing such extreme temperatures, even allowing for man-made climate change – this small probability reflects a general pattern of current climate models under-predicting extreme weather events.[14] But for the 'natural' model, the probability P_{NAT} of observing such an extreme result was estimated to be 0.023% (90% interval 0.018% to 0.030%), around 1 in 4,000. The Met Office warns that these estimates are based on numerous assumptions and should not be taken too literally, and declines to calculate the relative risk for such an extreme event associated with human influence. If the Met Office had been less cautious, it could have concluded that the relative risk P_{HUM}/P_{NAT} would be around 100, meaning that human influence made such an extreme event around 100 times more likely.

There is another vital lesson from Figure 10.2. The graph demonstrates that an apparently fairly small increase in average temperature, from around 12°C to 14°C, which we would probably not feel if it happened over an hour, leads

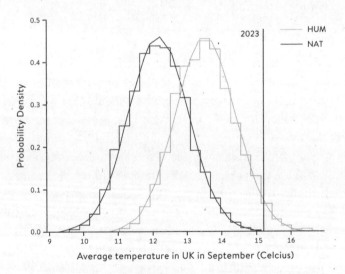

Figure 10.2
Probability distributions for average UK daily temperature in September 2023, generated from models assuming only natural variability (NAT – dark line), and human influences (HUM – pale line). The observed 2023 average temperature is shown as a solid black line and is implausibly high under natural variability.

to a disproportionate change in 'tail areas' and so massively increases the risks of extreme events. This simple graph brings home why trying to keep global warming below 2°C is such an important objective.

————————

Attribution studies have become more widespread, and are often carried out soon after, or even during, the event under scrutiny. This can lead to problems. In 2018, as Hurricane Florence was approaching the US coastline, researchers claimed that because of climate change the rainfall over the Carolinas would be increased by over 50%, and the hurricane would be about 80km larger. This was followed by headlines such as 'How global warming is turbocharging monster storms like Hurricane Florence'.[15] But two years later the same researchers acknowledged that a revised analysis estimated that climate change had made ±5% change in rainfall, and the hurricane was only about 9 kilometres (±6km) wider.[16] This retraction of their previous claims received little publicity, but demonstrated that such attribution studies should be carried out very carefully, allowing for the uncertainty about both natural climate variability and the outputs from climate models.

Unlike the caution expressed by the UK Meteorological Office, many attribution studies estimate relative risks associated with human-influenced climate change; for example, a relative risk of around $RR = 3.5$ was assessed for the record May temperatures in South Korea in 2017,[17] meaning such extreme temperatures were 3.5 times more likely to occur assuming human-influenced climate than assuming natural climate only. We saw earlier that in assessing causes of cancer we might calculate the *attributable fraction*, as the proportion of cases that would have been avoided had they not been subject to a particular exposure such as smoking, where the attributable fraction is given by $1 - \frac{1}{RR}$. In the climate context this is known as the *fraction of attributable risk* (FAR), and the FAR of the 2017 South Korea event was estimated to be $1 - \frac{1}{3.5} = 0.72$. Attribution studies for weather events currently steer clear of referring to this as a 'probability of causation', although it will be interesting to see if this starts to feature in future litigation around climate change.

Some climate scientists have taken the methods of attribution studies for specific events and, by considering the whole modern development of climate as a single 'event', applied them to climate change in general.[18] This leads to an estimated overall 'probability of causation' for human influence on temperature change of 0.9999 – providing an alternative quantitative basis for the IPCC's 2021 judgement that this was 'unequivocal'.

Nevertheless, in spite of the possible attraction of assessing a probability of causation, any estimate coming from an attribution study must rest on a wide range of assumptions – indeed, some measure of confidence in the analysis might be welcome. And this is for well-defined weather events, with reasonable understanding of the underlying physical processes. When we come to even more complex areas, for example legal cases claiming that social media companies have been responsible for harming the mental health of specific individuals, any attempts to quantify the uncertainty about causality will be even more fraught. But, as we will now see, there is a legal precedent in calculating 'probabilities of causation' directly from scientific evidence.

———————

We all have an interest in attributing ill-health to possible causes. Why have I got this headache? Who infected me with SARS-CoV-2? Did a vaccine cause my child's illness? Generally we just have suspicions, but some situations demand more formal methods. For example, when assessing whether a drug has led to an adverse event, an algorithm[19] is often used to classify reactions into *definite*, *probable*, *possible* or *doubtful* causation, although, like the IARC classification for carcinogenicity, this does not really express a probability but rather an informal strength of evidence for causality.

Of course, assessment of causality really reaches centre stage when there is a legal claim for damages, when a potentially life-changing decision must be made.

> How does a court decide whether someone has been harmed by their work environment?

John Cookson worked for a pharmaceutical corporation, later known as Novartis, for nearly thirty years, including a period in dye-stuffs production in Grimsby on the east coast of England. After retiring he developed bladder cancer, and in 2001 he sued his previous employer, claiming to have been exposed to 'aromatic amines' involved in the manufacture of dyes. It was accepted by all parties that these are associated with a raised risk of cancer. The court needed to decide whether, or with what probability, his cancer was caused by his exposure at work.

It's impossible to biologically identify where his cancer comes from. And we cannot rerun history and observe the counterfactual world in which Mr Cookson never goes near the dyes and see whether he would have developed cancer anyway. So to get at a reasonable probability that the exposure caused his *individual* cancer, we need to use the epidemiological evidence from *populations*. Essentially, the ideas of general causation (effects of causes) are used to attribute responsibility in a specific case (the causes of effects).

We have already come across the attributable or excess fraction associated with an exposure. For example, 93% of cases of lung cancer in female Norwegian smokers were attributable to their smoking. It is then a simple, but major step to go from this statement about populations to claim that if a particular smoker develops lung cancer, there is 93% probability that their lung cancer was caused by their smoking. If we are willing to make this jump, then this provides a way of estimating our probability of causation in an individual – just use the attributable fraction in the population. In general this will not be 1, as not every case of the disease was caused by the exposure. The legal issue is then whether the specific probability of causation is large enough to justify compensation.

While criminal convictions require evidence 'beyond reasonable doubt', which lawyers are reluctant to quantify, civil cases are decided on the 'preponderance of evidence' (US) or the 'balance of probabilities' (UK). This is often interpreted as meaning more than a 50% probability, and therefore a probability of causation greater than ½ should win the case for the plaintiff who is making the claim of harm.

There is a remarkably straightforward way of determining whether our probability of causation is greater than 50%; check if the relative risk is greater than 2.* The reasoning goes as follows; if an exposure more than doubles the risk of an adverse event, then more than half of cases can be attributed to the exposure, and so in any particular instance our probability of causation is greater than 50%.

Mr Cookson's case was made more complicated by his having been a moderate smoker for many years, although he had given up nearly fifteen years before his cancer was diagnosed.[20] It was accepted that both occupational exposure to various carcinogens and cigarette smoke were capable of causing bladder cancer, but Cookson's legal team argued that the relative risk

---

* Remember that if RR is the relative risk, the attributable or excess fraction is $AF = 1 - \frac{1}{RR}$. Therefore, if $RR > 2$, then $AF > \frac{1}{2}$.

associated with the chemicals was greater than 2, over and above the risk from smoking, and the judge in a 2007 appeal concluded that 'In terms of risk, if the occupational exposure more than doubles the risk due to smoking, it must, as a matter of logic, be probable that the disease was caused by the former.' John Cookson got his compensation.

––––––––

The 'doubling the risk' rule is written into US law, with the Energy Employees Occupational Illness Compensation Program Act of 2000[21] stating that, if an employee develops cancer after exposure to radiation at work, then compensation is payable if the assessed probability of causation is greater than 50%, and that this is determined by the relative risk being greater than 2. As we saw in the smoking example, limited epidemiological evidence means there will be uncertainty about the relative risk, which carries over into uncertainty about the probability of causality. Perhaps surprisingly, the US law states that the *upper* limit of a 98% confidence interval for the probability of causation determines whether the threshold of 50% is reached – this means that compensation is paid unless there is high confidence that the probability of causality is less than 50%.

In some circumstances relative risks less than 2 could still mean a probability of causation is greater than 50%. If exposure accelerates the condition, so that some of the cases would have developed the disease anyway if they had not been exposed, then this will not be reflected in the epidemiological relative risk.[22] This means the attributable fraction is only a lower bound on the probability of causation, where 'causation' now includes the harm of having the disease earlier than otherwise. So demanding relative risks greater than 2 may be too stringent.

All this assumes that the relative risk relevant to the specific case can be accurately assessed. The US Reference Manual on Scientific Evidence[23] for courts is suitably cautious about using the 'doubling the risk' rule, warning of the need for good evidence for the estimated relative risk, that the plaintiff should be similar to the study subjects, and that the exposure does not accelerate the disease and operates independently of other potential causes.

––––––––

A crime has been committed, someone has been accused, and they are in a criminal court of law. Although not the standard interpretation, this is essentially an attribution study, with two hypotheses, guilty or not guilty, being compared on the basis of evidence, much of it forensic. When assessing the 'guilt' for extreme weather events, researchers assess the probabilities of what was observed under two hypotheses for climate, natural or human-influenced, and may calculate their ratio. Similarly, the 'probative value' of forensic evidence is best determined by the likelihood ratio.

Going back to Chapter 7, remember the likelihood ratio expresses how much more likely an event B was to occur were A to be true, compared to if A were false, and so it essentially summarizes the information that B provides about A. In criminal law, the likelihood ratio takes the form

$$\text{likelihood ratio} = \frac{\text{Pr}\left(\text{Evidence} \mid \text{Prosecution proposition}\right)}{\text{Pr}\left(\text{Evidence} \mid \text{Defence proposition}\right)},$$

where the prosecution proposition may, for example, be the presence of a suspect at the scene of a crime, and the defence proposition would be that the suspect was not present.

Suppose the evidence comprises a DNA profile found at the scene of the crime, which exactly matches the suspect. The prosecution proposition is that the DNA is from the suspect, and so Pr(Evidence | Prosecution proposition) = 1, and so the likelihood ratio is

$$\text{likelihood ratio} = \frac{1}{\text{Pr}\left(\text{DNA profile} \mid \text{someone other than suspect}\right)}.$$

The probability of a particular DNA profile occurring from an unknown individual is called the 'random match probability' and is estimated using various assumptions about the frequency of particular elements of the profile in the population, although the exact values may be contested due to complications of the DNA sample. Typical DNA likelihood ratios are in the millions, or even, with a full profile, billions.[24]

Table 10.1 shows the recommended way that likelihood ratios should be reported in court[25] in the UK, similar to the translations between words and numbers described in Chapter 2. For example, in the analysis in Chapter 7, the Metropolitan Police were essentially assuming a positive live facial recognition identification has a likelihood ratio of around 700, which would be translated into 'moderately strong support' for the individual to be on the watchlist.

In the UK there is a 'cap': the highest likelihood ratio that can be reported is 1 billion (1,000,000,000).[26]

Value of likelihood ratio	Verbal equivalent
1–10	Weak support for proposition
10–100	Moderate support
100–1,000	Moderately strong support
1,000–10,000	Strong support
10,000–1,000,000	Very strong
1,000,000 and above	Extremely strong

Table 10.1
Recommended verbal interpretations of likelihood ratios in legal proceedings in the UK.

———

Going back to the live facial recognition example in Chapter 7, remember that while the system was claimed to have a *false alert rate* of 1 in 1,000, under plausible circumstances 59% of people picked up by the system would turn out to be false positive identifications. There is clearly concern about the potential confusion between

Probability of being picked out by the system, given you are not on the watchlist = 1 in 1,000 = 0.1%

with the 'inverse' probability

Probability of not being on the watchlist, given you are picked out by the system $= {}^{10}\!/_{17} = 59\%.$

The confusion between these two conditional probabilities even has a name: the **prosecutor's fallacy**. This arises through a common courtroom error. After DNA found at the scene of a crime matches that of a suspect, a reasonable statement about the random-match probability may be

If the suspect was not at the scene and someone else left the DNA, there is only a one in a million chance of having this degree of match

but instead the random-match probability may be wrongly interpreted as meaning

> With this degree of DNA match, there is only a one in a million chance of the suspect not being at the scene.

This can also be thought of as confusing the likelihood ratio (1,000,000) with the posterior odds that the prosecution is correct. Written so bluntly, it may be surprising that such an error can be made – it is like confusing 'most Popes are Catholics' with 'most Catholics are Popes'. But, as we shall see later, such misunderstandings not only happen but can have tragic consequences.

It's not just the prosecutors who have a fallacy. Suppose someone is suspected of leaving DNA at the scene of a crime, and the random match probability is 1 in 5 million. This provides a likelihood ratio of 5 million, or 'extremely strong evidence' in favour of the prosecution's argument. But the defence then points out that there are 60 million other people in the UK, and therefore around 12 others who match the suspect's DNA. So there is only a 1 in 13 chance that the suspect was at the scene. This is sometimes called the 'defence fallacy', since it assumes everyone in the UK was equally capable of being at the scene of the crime.

This is a form of Bayesian argument, in that essentially the suspect is given initial prior odds of 1 to 60 million of being at the scene, which is then multiplied by a likelihood ratio of 5 million, to give posterior odds of 1 to 12. So this is only a 'fallacy' in the sense that the prior probability should be based on the number of individuals who could potentially have been at the scene of the crime, and it does not take into account other corroborating evidence.

————

There has been a lot of discussion of the sad case of Sally Clark, but much of it has missed the main point. She was a solicitor who had two babies, aged seven weeks and eleven weeks, die suddenly and unexpectedly a year apart. She was then convicted of their murder, and given a life sentence in 1999. At her trial, Professor Sir Roy Meadow, a retired paediatrician and an expert on Sudden Infant Death Syndrome (SIDS), claimed that the chance of a child in such a family dying from SIDS was around 1 in 8,543, and that the chance of two babies dying from SIDS in the same family was therefore $\frac{1}{8,543} \times \frac{1}{8,543}$ or about 1 in 73 million, which would mean we would expect this to occur about once in every hundred years in England and Wales. He described these odds as similar to backing a long-odds horse at 80 to 1 for the Grand National and its winning four years in a row, and that such an event is 'very, very, very unlikely'.[27]

There are two problems with this reasoning. First, the multiplication of the probabilities is only valid if the events are independent, and it is known that deaths from SIDS tend to cluster in families, possibly through genetic associations. So 1 in 73 million is far too small a probability.

But this is not the most important issue. As statistician Philip Dawid has pointed out,[28] 'if background evidence of double-death-rates due to SIDS (or other natural causes) is relevant, then surely so too should be background evidence of double-death-rates due to murder.' Essentially, while Meadow was right that two deaths due to SIDS is extremely rare, it is also 'very, very, very unlikely' that a mother will murder two of her young children. Again, this is more of a Bayesian argument, since we are taking into account a very low prior probability that someone will commit such a crime.

The case was first appealed in 2000, but testimony from expert statisticians such as Professor Dawid was disallowed on the grounds that 'It's hardly rocket science, is it?'[29] Finally, at a second appeal in 2003, new evidence was presented about bacterial infection in one of her sons that had been previously withheld, and Clark was freed.[30] The legal judgement heavily criticized the pathologist, but also said that Meadow's evidence should never have been put before the jury in the way it was, and that 'the graphic reference by Professor Meadow to the chances of backing long-odds winners in the Grand National year after year may have had a major effect on their thinking.' Although the new pathology evidence led to Clark's appeal being granted, the judges concluded that they would 'in all probability, have considered the statistical evidence provided quite a distinct basis upon which the appeal had to be allowed'. The successful appeal led to re-examination of other cases in which Meadow had testified, and three women were subsequently cleared of murdering their children. Sadly, after having spent four years in prison before she was exonerated, Clark died through acute alcohol poisoning in 2007.

The case of Australian Kathryn Folbigg is, if possible, even more tragic. All her four children died as babies, in 1989, 1991, 1993 and 1999. Although there was no direct evidence of her harming the children, she was convicted in 2003 of smothering the children and given a forty-year sentence, largely on the basis of the supposed improbability of all the children having died from natural causes. Roy Meadow was cited in support of the deaths being beyond coincidence. After a long campaign, she was finally freed in 2023 due to new evidence that her children carried some very rare gene mutations which predisposed them to sudden cardiac death. She had spent twenty years in prison.

As we saw in Chapter 4 on coincidences, courts need to take great care before concluding the events are so unlikely that they 'cannot be just chance'. First, because given enough opportunities, even apparently rare things will happen. Secondly, because there may be common factors behind the events that dramatically increase the chance of them occurring together. Finally, because we also need to take into account the rareness of alternative, criminal explanations.

All this can be clarified using formal analysis of likelihood ratios and Bayesian thinking. But although evidence-based likelihood ratios are permitted in UK legal proceedings, the Court of Appeal for England and Wales has ruled that Bayes' theorem should not be formally used in court to combine and weigh evidence.[31] This is apparently best done by the jury using their human judgement.

––––––––

After covering causes of cancer, attribution of climate change and extreme weather events, civil court cases for damages, and criminal cases relying on forensic science, it is time to reflect on what they all tell us about uncertainty and causality.

The first lesson is that assessing causality is difficult. There is a lot of uncertainty, and it is not straightforward to express it in terms of probabilities. Second, in some circumstances it *is* possible to come up with a probability of causality, but we need to distinguish two different types of question. In many circumstances we cannot directly observe a causal link, whether between human activity and extreme weather, or between exposure to chemicals and getting a cancer. We cannot absolutely prove the chain of causation, and so have to deal with associations, and infer causality. In contrast, in criminal cases we directly examine causal hypotheses about why events occurred, which could in principle be decided with certainty if we had the right evidence. Which means that in theory a 'probability of guilt' could be calculated (although this would be inadmissible in a UK court).

Finally, there is often a confusion between probability and strength of evidence. A famous legal question, first posed in 1971,[32] may clarify the issue:

The claimant is negligently run down by a blue bus. The only issue is whether the bus was operated by the defendant, who operates 80% of all

the blue buses in town. If this is the only evidence at the trial, is it sufficient to prove the claimant's case to the civil standard of proof?

In a recent informal poll of lawyers at a conference,[33] around two thirds said
this was sufficient to conclude on the balance of probabilities that the defendant's buses were responsible, while the other third disagreed, presumably
feeling that although the probability was greater than 50%, only circumstantial evidence had been presented.* From a Bayesian perspective, we would
say that we have 80% as a prior probability, but with no evidence specific
to this case.

In his 1921 *Treatise on Probability*, John Maynard Keynes wrote:

> As the relevant evidence at our disposal increases, the magnitude of the
> probability of the argument may either decrease or increase, according as
> the new knowledge strengthens the unfavourable or the favourable evi-
> dence; but something seems to have increased in either case – we have a
> more substantial basis upon which to rest our conclusion.[34]

This reflects our discussion in Chapter 9, where direct uncertainty, preferably
expressed in terms of probability, is clearly distinguished from indirect uncertainty, which is about strength and relevance of evidence. In particular, we identified that people were most unwilling to make decisions based on the current
probabilities if there were major information gaps, where potentially available
evidence could dramatically change their current beliefs. UK Supreme Court
judge Lord Leggatt has made the same point about the bus example, saying,
'first, too much relevant information is missing and, second, the available information is insufficiently specific'.[35]

In general, the legal system finds it challenging, to say the least, to use statistical and epidemiological evidence in its deliberations.[36] Perhaps a clearer
distinction between probability and strength of evidence would help in this
and many other areas.

---

* Although it would be interesting to know how many would still disagree if, say, the defendant operated 999
out of 1,000 blue buses.

Summary

- Uncertainty about general causality – the effects of causes – is usually expressed as a qualitative judgement.
- In this context, terms such as 'probable cause' are based on strength of evidence and cannot be interpreted probabilistically.
- Uncertainty about attribution, the causes of specific events, may in some circumstances be quantified as a 'probability of causation'.
- Climate attribution studies estimate the relative risk of extreme weather events associated with human influences on climate, although there is substantial uncertainty about these assessments.
- In civil legal cases, the 'doubling the risk' rule is sometimes used as a basis for claiming the probability of causation is greater than 50%, although this may be too stringent.
- In criminal cases, the value of forensic evidence is best summarized by a likelihood ratio comparing the prosecution and defence hypotheses.
- All these areas would benefit from a clearer distinction between probability and strength and relevance of evidence.

Predicting the Future

We don't know what is going to happen in the future and, unless we are blessed with some magical skills, we cannot know. Uncertainty about the future is therefore fundamentally different from the epistemic uncertainty of the preceding chapters, where our current ignorance could, at least in principle, be eliminated by more knowledge. When it comes to wondering what will happen, we just have to wait and see what turns up.

This basic indeterminacy has not prevented people from seeking answers to the unanswerable question – what is going to happen to me? Fortune-tellers and oracles have thrived throughout history and many have made use of randomness as a vehicle for their prophecies. The aim of such *cleromancy* is to mirror the patterns of the future by the patterns of, for example, the cast of the yarrow stalks for the *I Ching*, the selection of Tarot cards, or the shapes made by tea leaves, notably in Professor Trelawney's class in *Harry Potter and the Prisoner of Azkaban*.[1] Of course, as statistician David Hand has observed, it is important that the soothsayer uses obscure language and makes their prognostications as numerous and ambiguous as possible.[2]

In complete contrast, the Scientific Revolution ushered in a new, more rigorous, era. Scientists for the first time were able to make highly specific predictions using transparent methods based on mathematical equations representing physical laws. For example, using Newton's model for planetary motion, Edmond Halley calculated the orbits for twenty-four historical comets but noticed that three of them seemed remarkably similar. By assuming they were all the same comet following an elliptical orbit, in 1705 he predicted it would next appear in 1758. Sadly he died in 1742, before he could witness his comet returning on schedule, but, as we shall soon see, not before contributing fundamental ideas to predicting how long people might live.

Isaac Newton might appear to epitomize scientific rationality, but in fact he was obsessed with alchemy (almost killing himself in an explosion in his laboratory at Trinity College Cambridge) and interpreting numbers in the Bible. He was an 'Arian' – one who did not believe that Christ was divine – and thought the established Church was corrupt. He had to keep these heretical beliefs secret, but privately he used biblical references to estimate when the world as we know it would end and Christ would come again to form a new global kingdom of peace.[3] He concluded that this would be around 2060,[*] coincidentally just before the next appearance of Halley's Comet in 2061.

———

Newton's laws are examples of the deterministic physical principles which still form the basis for predictions in much of human activity, whether planetary motion, weather, climate, landing a spacecraft on an asteroid or hitting a target with a missile. But even the simplest flipped coin defies deterministic prediction (unless you are Persi Diaconis), and so we need to acknowledge uncertainty, which, in some circumstances, can be expressed as a probability. While the sort of forecasting competitions we saw in Chapter 2 may use people's subjective judgements, we generally turn to data-based statistical models to produce numerical assessments of uncertainty about the future.

The examples in this chapter appear in order of how far ahead we want to predict; there is a big difference between predicting weather *vs* predicting climate, football results *vs* eventual sporting records, whether I will survive next year *vs* how long people will live in the future, next year's inflation *vs* inflation in forty years' time, let alone the future of humanity. Shorter-term predictions can generally be made with more confidence as things can be considered stable, while longer-term predictions should be increasingly dominated by deeper uncertainty about how the world will develop. But, as always, any analysis depends crucially on assumptions – even in the extremely short term, you are not going to make a good assessment of your

---

* Newton based his predictions on three references in the Book of Revelations to 1,260 days (42 months), for example, 'And I will give power unto my two witnesses, and they shall prophesy a thousand two hundred and threescore days, clothed in sackcloth' (Revelations 11.3, King James Version). He translated this to 1,260 years, and added it to the date of the formation of 'the Pope's supremacy', which Newton set at 800. This gave 2060 as the date of the end of the corrupt Trinitarian Church.

probability for a coin flip coming up tails if someone has passed you a two-headed coin.

Predicting next week's football results*

It was 22 May 2009 and the English Premier League had one match left to play, with West Bromwich Albion at the bottom of the league with 31 points, and Manchester United at the top with 87. I had been asked by the BBC radio programme *More or Less* to make some predictions about the final matches, and so used a basic statistical model to assess the chances of any particular result for all the matches to be played at the weekend – much more sophisticated analysis is now used by both sports-betting companies and punters.

Table 11.1 shows the working for Wigan *vs* Portsmouth, then at positions 12 and 14 in the league of twenty teams. Wigan had 42 points and had scored 33 goals. The average number of goals scored up to then was 46, and so we can estimate Wigan's 'strength in attack' as $33/46 = 0.72$, meaning they have only scored 72% of the average number of goals. Similarly, Portsmouth have conceded 56 goals compared to an average of 46 (goals conceded must match the number of goals scored), leading to their estimated 'weakness in defence' of $56/46 = 1.22$, meaning Portsmouth have let in 22% more goals than average.

Together, these allow us to estimate how many goals we expect Wigan, the team playing on their home pitch, will score. We start from a baseline expectation of 1.40, the average number of goals scored by a home team. Then we adjust this by Wigan's attack strength of 0.72, and Portsmouth's defence weakness of 1.22, to get $1.40 \times 0.72 \times 1.22 = 1.22$ goals.

Similarly, Portsmouth starts with a baseline of 1.08, the average number of goals scored by an away team, which is adjusted by their attack strength of 0.83 and Wigan's defence weakness of 0.98, to give $1.08 \times 0.83 \times 0.98 = 0.87$ goals. But, just like nobody has 2.4 children, nobody scores 0.87 goals – this is only an expected value, a theoretical average if the match were played again and again (heaven forbid). To get the probability for each specific number of goals, it is reasonable to assume a Poisson distribution, arising naturally from the large number of possible low-probability opportunities for scoring. This

---

* I apologize to those who don't care about football for featuring the subject again. It probably is no consolation that I also have no interest in the sport, but this does provide a good example of a moderately complex model leading to short-term probabilistic predictions.

	Points	Goals for f	Strength in attack $f/46$	Goals against a	Weakness in defence $a/46$	Baseline expectation (home or away)	Expected number of goals
Wigan	42	33	0.72	45	0.98	1.40	$1.40 \times 0.72 \times 1.22 = 1.22$
Portsmouth	41	38	0.83	56	1.22	1.08	$1.08 \times 0.83 \times 0.98 = 0.87$

	% probability of scoring x goals					
x	0	1	2	3	4	5
Wigan	29%	36%	22%	9%	3%	0.7%
Portsmouth	42%	37%	16%	5%	1%	0.2%

Table 11.1
The model for predicting the result of the Wigan–Portsmouth Premier League match on 24 May 2009. The probabilities for scoring each number of goals are derived from Poisson distributions with means 1.22 and 0.87.

gives the probability distributions shown in Table 11.1; for example, we assess a 37% probability that Portsmouth will score precisely 1 goal.

To assess our probability of an actual result of the whole game, we might assume the goals scored by each team are independent, in the sense that if we knew how many Wigan scored, it would not give us any additional information about Portsmouth's performance. This is a strong assumption, but it means that we can find, for example, our probability of a 1–0 result, which is the most likely outcome, by multiplying 36% by 42% to get 15% – so even the most likely result is still not very likely.

In fact, there tends to be some correlation between teams' results, in that matches have some tendency to be either high- or low-scoring. Special software that allows for such correlations provided estimated probabilities for every combination of goals and led to the assessments for home win/draw/away win in Table 11.2.

The 'most likely' goal combinations were announced by James Alexander

Home team	Away team	Probability of home win (%)	Probability of draw (%)	Probability of away win (%)	Brier penalty score
Arsenal	Stoke	**72**	19	10	0.12
Aston Villa	Newcastle	**62**	21	17	0.22
Blackburn	West Brom	54	**23**	23	0.94
Fulham	Everton	35	**35**	30	0.64
Hull	Man United	9	19	**72**	0.12
Liverpool	Tottenham	**72**	20	9	0.13
Man City	Bolton	**59**	22	19	0.25
Sunderland	Chelsea	10	25	**65**	0.20
West Ham	Middlesbrough	**57**	28	15	0.29
Wigan	Portsmouth	**44**	32	25	0.48
				Average	**0.34**

Table 11.2
Assessed probabilities of home win/draw/away win for all Premier
League matches played on Sunday, 24 May 2009. Actual result shown in
bold, with accompanying Brier penalty score.

Gordon (who used to read the actual results out on the BBC) on the *More or
Less* broadcast on 22 May 2009[4] and, somewhat to our consternation, were
reported as definite predictions without any qualifying probabilities.[5] We
spent the weekend in nervous anticipation.

When the real results came in on 24 May, Table 11.2 shows we got 9/10 'correct' in terms of the most likely predictions of home win/draw/away win, plus
two exact scores , including Wigan's 1 – 0 win over Portsmouth! This was grat-
ifying, especially as Mark Lawrenson, the official BBC football expert, only got
seven correct results and only one exact score.[6]

All very good, but we know that this is not how we should assess probabilistic
forecasts. For our quiz in Chapter 2 we adapted the Brier scoring rule, developed in the field of weather forecasting, to assess how accurate our probabilities were. The Brier penalty score is used in its original form in Table 11.2, where

high scores correspond to poor predictions: a Brier score of 0 corresponds to a perfect prediction and a score of 2 to a useless prediction that put 100% probability on an outcome that did not occur.* The average Brier penalty was 0.34.[7]

It is helpful to have something to compare this score with, say a prediction that did not use any knowledge of the individual teams – essentially one with 'no skill'. A default prediction for all the matches could have been the probabilities 0.45, 0.26 and 0.29, since these are the proportions of home wins, draws and away wins throughout the season. This would have given a Brier penalty of 0.59, rather worse than the average score of 0.34 that we achieved, and so our model has allowed us to reduce our penalty by 0.59 – 0.34 = 0.25 compared to a 'no-skill' prediction. The percentage reduction is $0.25/0.59$ = 43%, and this is known as the Brier Skill Score (BSS), where a skill score of 0% indicates predictions that are essentially no better than chance, and a skill score of 100% indicates perfect predictions.

It also turns out that our Brier penalty was slightly less than we would have reasonably expected it to be, had the probabilities in Table 11.2 been the 'true' chances of each result, and this confirmed our impression that we were rather lucky. And so were some people who, against our recommendations, placed bets on our choices.

Unfortunately our luck did not hold out. I tried the exercise again the following year and did badly. I should have stopped while I was ahead.

Predicting the weather for the next week

On 15 October 1987, well-respected BBC weather forecaster Michael Fish said during a broadcast 'a woman rang the BBC and said she had heard that there was a hurricane on the way. Well, if you are watching, don't worry, there isn't.' Unfortunately for his reputation, there really was a hurricane on the way, which during that night killed 22 people, blew down 15 million trees, and caused more than £2 billion in damage.

Weather is a classic example of a chaotic system, where complex non-linear processes can lead to extreme sensitivity to initial conditions. Forecasts were traditionally based on human judgements derived from observations made at a network of weather stations, but by the 1950s computers allowed

---

* If p_H, p_D and p_A are the probabilities given to a home win, a draw and an away win respectively, and say the actual result is a home win, then the Brier penalty is $(1 - p_H)^2 + p_D^2 + p_A^2$. So it is the sum of the squares of the errors. So, for example, the Brier penalty score for Wigan vs Portsmouth was $(1 - 0.44)^2 + 0.32^2 + 0.25^2 = 0.48$.

numerical predictions based on models that represent the movements of the atmosphere as mathematical equations operating on a grid. These produced a single deterministic forecast of the type available to Michael Fish in 1987, without any measure of uncertainty.

Spurred on by the debacle of the 1987 Great Storm, Tim Palmer and his team at the European Centre for Medium Range Weather Forecasts (ECMWF) started adapting the Monte Carlo approach to these big-weather models. By running the model from fifty different sets of initial conditions, essentially fifty different 'possible futures' could be examined – this collection of forecasts is known as an **ensemble**. When they retrospectively examined the data from October 1987, and started the model at fifty different perturbations of the initial conditions for midday on 13 October, they found that sixty-six hours later (early on the 16th), many of the fifty members of the ensemble showed a substantial depression over the south of the UK, with over 30% showing hurricane-force winds at some point.[8] Michael Fish would not have been so confident if he had had access to this intelligence.

Ensembles appear to provide a natural means of assessing uncertainty – if 20 out of 50 members of the ensemble show rain at a particular time and place, then announce a 40% probability of rain. But, to go back to a running theme in this book, probability assessments should be well calibrated, so that of the times when a weather forecast declares a 40% probability of rain, then it should rain in about 40% of cases. Palmer describes the extreme challenge of producing calibrated probabilities from an ensemble, since just randomly perturbing the initial conditions does not fully explore the possibilities, with the ensemble being too tightly clustered and claims being too confident. Instead, the perturbations need to be deliberately focused on directions in which the atmosphere is least stable.

Ensembles started to be used by the ECMWF in 1992 and have now become a standard way of producing probabilistic forecasts. When it comes to assessing their quality, it is natural to go back to Glenn Brier's work on scoring rules, developed when probabilities were just subjective judgements. Just as for our football predictions above, the 'skill' of a forecast system is measured by the improvement in penalty score over a baseline 'skill-less' forecaster. For football, we used the long-run average proportions of each type of result as the baseline, while in weather forecasting they use the predictions we would make from the long-run climate, for example on what proportion of days we would expect it to rain at this time of year.

ECMWF report that their current Brier Skill Score for predicting rainfall in Europe – the % improvement over just predicting using climate data – is around 40% for two days ahead, and 20% for seven days ahead.[9] This may not sound great, but it has been steadily improving as computer power allows finer granularity, and it would be unreasonable to expect the near-perfect prediction needed to get skill scores nearer 100%.

Nevertheless, there is competition from a radically different modelling approach based on *deep learning*, often labelled as artificial intelligence (AI). This makes no effort to represent the underlying physical processes by equations, and instead builds a complex multilayer network relating a range of weather variables. For example, the 2023 version of GraphCast[10] uses thirty-nine years of historical data to train a network with 37 million parameters (small by modern machine-learning standards). The predictions, essentially based on a black box of statistical associations rather than a causal model, are claimed to have substantially better skill than standard systems when making deterministic predictions. The 2023 version of GraphCast does not handle uncertainty, although the researchers state this is a crucial next step.

Things have come a long way since Michael Fish's unfortunate performance.

Predicting Covid weeks and months ahead

During the pandemic we all became familiar with images of projections (note, not 'predictions') into the future of the numbers of Covid-19 cases, hospitalizations and deaths. Just as when estimating R (Chapter 8), multiple teams constructed models using a range of different approaches, from curve-fitting to complex deterministic models for the whole population. Often the projections were shown with (generally very wide) bands of uncertainty, although these were generally ignored by media commentators.

Unlike weather forecasting, pandemic models are not particularly sensitive to initial conditions, but they are extremely sensitive both to choices of the model structure, as we've already seen when estimating R, and also to all the assumptions that have to be made about the characteristics of the latest variant of the virus, and the effectiveness of vaccines and non-pharmaceutical measures such as social distancing. An even more fundamental difference is that, again unlike the weather, the progress of a pandemic is strongly influenced by people's behaviour, and this behaviour may itself be influenced by the publicized projections, creating a feedback loop. The projections are therefore extremely sensitive to the very factor that is known least about, which is human behaviour in the future. This sensitivity made it essential that the projections were reported as possible scenarios, and not predictions or forecasts, although this is often how they were interpreted.

The potential for misunderstanding was made even worse as the modellers in the UK were generally responding to government requests for projections under particular assumptions, including what are known as reasonable worst-case planning scenarios (RWCS). These are necessarily pessimistic, so it is unsurprising that events often turned out somewhat better than the projections. This led to some additional scepticism.

The problem is that if modellers genuinely cover all the uncertainties about the parameters, particularly changes in human behaviour, and allow for different choices of model, then the resulting uncertainty intervals for anything more than a few days or weeks will be very wide indeed, covering essentially all possibilities. And then the policymakers may well ask – what's the good of that? To which the modellers might answer: we have been given an impossible task – we cannot say with any confidence what will happen in the longer-term future.

Sometimes it might be better for analysts to refuse to try to answer an unanswerable question, and just say – it depends.

Predicting the economy a few years ahead

In May 2018, the Bank of England released its quarterly Inflation Report, which contained the projections for annual change in GDP for the next three years, shown in Figure 11.1(a).[11]

Things turned out rather differently to the projections. Figure 11.1(b) shows that, following the Covid-19 pandemic, the first quarter of 2020 saw GDP temporarily plummet, with an annual reduction of 25%, way off the scale of the chart from May 2018. On the face of it, this is a huge failure of the projections shown in Figure 11.1(a), so should the Bank of England be considered 'wrong'? This all depends on their expression of uncertainty.

The Monetary Policy Committee (MPC) of the Bank has used fan charts (with some redesigns) since the 1990s, designed to emphasize not only the uncertainty about the future but also about the present and the past – Figure 11.1(a) and (b) show the current estimates of past growth by the Office for National Statistics (ONS), and the considerable uncertainty reflects potential revisions as more data comes in. For the future, the MPC says that 'If economic circumstances identical to today's were to prevail on 100 occasions, the MPC's best collective judgement is that the mature estimate of GDP

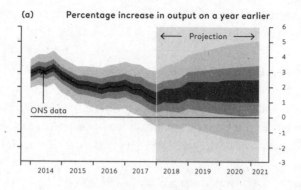

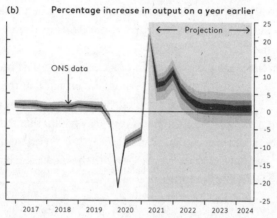

Figure 11.1
(a) The Bank of England's 'fan chart' showing projections for future growth and assessments of past growth, as issued in May 2018. The bands show the central 30%, 60% and 90% probability intervals. (b) The fan chart from August 2021, showing what happened after 2018. Note the change in vertical scale.

growth would lie within the darkest central band on only 30 of those occasions.' So the MPC's interpretation of probability is essentially the expected frequency of possible futures – as if the world were replayed a hundred times. As we saw in Chapter 3, this fits well with the understanding of Alan Turing, Richard Feynman and others.

The 'fan' is primarily based on a statistical model, with explicit assumptions about stability of asset purchases by the bank, and then adjusted for future uncertainty according to the subjective opinion of the MPC, itself based on the size of historical prediction errors and judgements about future risks. A smooth curve is then fitted to the central 90% of their distribution.[12] There are two notable features of the resulting charts:

1. The MPC does not give a central estimate for the future, presumably to avoid undue focus by commentators.
2. The tails of its probability distribution are not modelled, with the MPC saying, 'And on the remaining 10 out of 100 occasions GDP growth can fall anywhere outside the green area of the fan chart (grey area).'

Essentially, the MPC gives 10% probability to 'something else' occurring, which means it cannot be accused of being 'wrong' when it failed to predict the extreme consequences of either the 2008 financial crash or the 2020 Covid pandemic.

The projections are a combination of assumption-based modelling and subjective assessment – Type A and Type B uncertainty, in the language of metrology – while leaving room for unmodelled unknowns. They are powerful communication tools, although not everyone may realize that they leave open the possibility of a massive crisis. Fan charts are also used by the European Central Bank, but have never been popular with the media and have fallen out of favour with other central banks.*

Predicting our lifetimes

Halley may have his comet, but that's not why he is a hero to statisticians. In 1693 he published a paper in the Transactions of the Royal Society in London, in which he analysed data on the ages at which people had died in the city of Breslau (now Wrocław in Poland) between 1687 and 1691. The idea of 'climacteric' ages, in which people were at particular risk of dying, had been around for centuries – sixty-three years old was considered the 'grand climacteric', and so particularly dangerous. But nobody had actually looked at data, until Pastor Neumann of Breslau circulated his records around Europe. By looking at the number dying at each age, Halley not only put to rest the idea of climacteric

---

* Just as this book was being completed, the Bank of England published a review of their forecasting by Ben Bernanke, former chairman of the US Federal Reserve. He was critical of fan charts, and recommended they should be replaced by central projections from improved models, supplemented by alternative scenarios, verbal statements of uncertainty and a summary of past forecasting errors. While these are valuable additions, I personally feel it is unfortunate to lose a strong visual representation of modelled uncertainty.

years but, crucially, estimated the proportion dying at each year of age, of those who had survived until then – this then became known as the *force of mortality*, although we would now call this the *hazard*.* He therefore created the first proper *life table* showing the estimated probabilities of surviving until each age – it went up to eighty-four, an age he estimated a 2% chance of achieving.

A recent innovation in the late 1600s had been the sale of annuities as a way of raising money for the government, where for a single payment by a customer the government guaranteed to pay them a fixed sum each year for the rest of their life. The standard was to charge seven times the annual payment, so for example if someone wanted £100 a year for the rest of their life, they would pay the government £700. But this was the rate regardless of the age of the customer! So it could be a massive loss-maker if younger, healthier people bought annuities. It was a small but brilliant step for Halley to use his life table to set a minimum price for an annuity that should yield a profit to the government. Halley's 1693 paper showed that a customer aged twenty should be charged 12.8 times the annual payment, one aged fifty 10.9 times, and that the standard '7 times' rate should only be offered to people aged over sixty-five. Then, having started the annuity (and life insurance) industry as a sideline, Halley returned to his general scientific work, sorted out his comet, and as a final achievement died at eighty-six, two years beyond the end of his life table.

Figure 11.2 displays some of the information from the current life tables for the UK, specifically the estimated hazard rates (the proportion of survivors who die at each year of age) and the consequent distribution of ages at death, which is the proportion of births who are expected to die at each year of age, *assuming the current hazards apply for their whole lifetime.*

Figure 11.2 (a) shows the hazards increasing sharply with age, although the 'spike' just after birth is invisible on this scale. Males have consistently higher annual risks of death than females being roughly 50% higher between the ages of forty and eighty, meaning that, at each age, three males die for every two female deaths.

Better insight is gained by displaying this showing on a logarithmic scale, as in Figure 11.2 (b). This scale reveals the relatively high mortality in the first years of life, generally due to congenital diseases or problems around birth,

---

* You may remember we have already defined a 'hazard' as the potential to cause harm, rather than a numerical probability. Confusingly, in the world of survival analysis, the hazard is the precise opposite – it's the probability of a 'failure' in a short period of time. I am afraid this is typical of statisticians taking everyday words and giving them technical definitions – see 'expectation', 'odds', 'significance', 'likelihood', etc., etc.

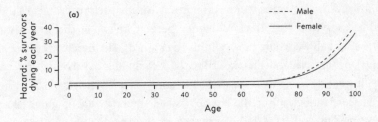

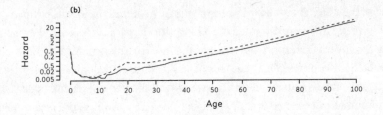

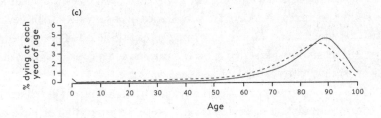

Figure 11.2
Hazard rates (annual probability of dying, given survival so far) on
(a) a linear and (b) a logarithmic scale; (c) the distribution of age
at death assuming current hazard rates remain throughout each
lifetime. UK life tables, 2018–20, for males and females.

with annual risk falling to a minimum within a few years – since only around 1
in 15,000 nine-year-olds die each year, this is possibly the safest that any group
of people have been in the entire history of humanity. After that the risk grows
inexorably – rash behaviour produces a sad rise, particularly in males, in late
teens and early twenties, and then the lines are fairly straight. This is *Gompertz's
Law*, first observed by Benjamin Gompertz in 1825,[*] and it means that the aver-

---

[*] This appears to be a counter-example of Stigler's Law of Eponymy, since, remarkably, it does seem to have
been discovered by Gompertz himself.

age annual risk of death increases at the same rate of about 9% per year, meaning it roughly doubles for every eight years increase in age.

Figure 11.2(c) shows the probability distribution of age at death expected for someone living with the current average hazards. The mean average of these distributions is the life expectancy at birth, and is currently eighty-three for females and seventy-nine for males, while the mode of the distributions shows the most common age at which people would be expected to die, which is eighty-nine for females and eighty-six for males – the skewness of the distribution produces this substantial difference between the average and the most common age at death.

But where are each of us going to be on this curve?

How long are you (or I) going to live?

You are not the only one who might be interested in the answer – it is just what actuaries ask when they price up annuities and life insurance, building on Halley's work from over 300 years ago. We could use the life-table data to form a distribution for the age of death among people achieving your age. But this would assume both that these risks apply to you, and that they will continue to apply to you in the future. Both assumptions are generally inappropriate.

First, the published hazards show, of all people who reach each birthday, the observed proportions who die in the subsequent year. This is a description of a population and can be thought of as the mean-average risks at each age. But much of the population's mortality risk is covered by those who are already ill, so the distribution of risk at each year of age will be highly skewed, with the mean average being pulled up by a small group of people at very much increased risk. This means that the risk of the 'average person' – the median – is substantially lower than the published hazard. It may seem paradoxical, but most people are at lower risk than average.

So what would be your personalized life table? We can think of the published ones as baselines, in which the hazards can be adjusted up or down according to your individual risk factors. This is known as the **proportional hazard model**, for example someone who smokes twenty cigarettes a day has around double the annual risk of death compared to a non-smoker, and this produces around eight to ten years' reduced life expectancy. In fact, a rough rule of thumb is that something that is associated with an increased hazard of 10% a year, such as sedentary behaviour equivalent to watching two hours of television a day,[13] reduces your life expectancy by around one year – we shall use this later when we consider the communication of chronic risks in Chapter 14.

There are increasing claims about personalized risk assessments based on genetics, summarized by so-called **polygenic risk scores**, but their importance may be exaggerated as they generally appear to add little over what can be gleaned from basic information about age, sex, lifestyle and family history.[14] For example, in 2019 England's then Health Secretary Matt Hancock was 'surprised and concerned' that a genetic profile estimated he had a 15% risk of developing prostate cancer by the age of seventy-five, and said, 'The truth is, this test may have saved my life.' But this was later widely ridiculed,[15] since Cancer Research UK reports that one in six UK males will be diagnosed with prostate cancer in their lifetime anyway. There may be greater value in personalized predictions for those who already have a disease. After my diagnosis with my own prostate cancer in 2016, I struggled to find good information on the survival for people like me – although a recent algorithm[16] would assess around a 77% 10-year survival up to 2026 (which I hope to be able to achieve).*

When we apply for life insurance and click the boxes asking about age, smoking, family history of disease, and so on, the insurance company's algorithm is applying these adjustments to your personalized life table. But however refined the analysis, it can never really quantify *your* risk (which does not really exist) – just what we would expect to happen to a group of people who ticked the same boxes as you did.

———

Another problem with using life tables as predictions is that they give what is known as the *period life expectancy*, which is based on assuming that current hazard rates will stay the same in the future. But if we want to assess how long we would expect a newborn to live, we need the *cohort life expectancy*, which builds in projections about how the hazards will develop.

Table 11.3 shows that girls born in England and Wales in 2020 are currently

---

* These numbers are essentially based on the experience of past patients who matched my profile. One issue is that as the relevant information gets more personalized, the pool of past similar patients gets smaller, and so although the risk assessments become less biased, their uncertainty grows. This is an example of what is known as the 'bias/variance' trade-off.

	Period life expectancy – assuming hazards at the year of birth hold throughout their lives	Cohort life expectancy – allowing for changes in hazards	% reaching 100
Females born in 2020	82.6	90.3	19%
Females born in 2045	85.5	92.7	27%
Males born in 2020	78.6	87.5	14%
Males born in 2045	82.4	90.2	21%

Table 11.3
Estimated period and cohort life expectancies for births in 2020 and 2045, England and Wales, with estimated percentage who will reach 100 years old. UK Office for National Statistics (ONS).[18]

on average expected to live to 90.3, and 19% to reach 100, while the next generation of girls born in 2045 have an estimated cohort life expectancy of nearly 93, with more than one in four getting to their hundredth birthday in 2145. Let's hope there is someone to look after each of them.

All these estimates are subject to both substantial variation and uncertainty. Life expectancies in areas in the UK with the highest levels of deprivation, such as Glasgow City, are around ten years lower than in richer areas such as South Cambridgeshire. And even before Covid, life expectancy was declining in nearly 20% of communities for females and in around 11% of communities for males.[17] There is also, of course, considerable uncertainty about what the future will bring, with mortality rates possibly being affected by climate change, pandemics, conflict, and so on.* The UK Continuous Mortality Investigations model used by the insurance and pensions industry allows users to put in their own assumptions, although their illustrative examples assume a continued annual reduction in mortality rates of 1.5%,[19] corresponding roughly to an extra two months of life expectancy each year.

So how long are we going to live? I can use myself as an example. I am

---

* Usually the ONS includes additional high and low scenarios to reflect a wider range of assumptions, but the most recent interim analyses only included a principal projection, with a warning about the uncertainty 'in setting long-term demographic assumptions following the coronavirus (COVID-19) pandemic'.

seventy at the time of writing, and according to the latest life tables for England, the period life expectancy for a male of that age is another fifteen years, taking me to eighty-five, with a 26% chance of reaching ninety, and 1% of celebrating my hundredth birthday. But that is just a baseline – I am reasonably fit for my age, don't smoke and am not (too) overweight, but on the other hand I still have prostate cancer. Let's be very optimistic and assume these factors cancel out and the published figures apply to me. If we assume that mortality rates will continue to reduce at the actuarial baseline of 1.5% each year, then my cohort life expectancy is another seventeen years, with a 34% chance of reaching ninety, and 5% of getting to a hundred.

At the moment, I feel that ninety seems more than enough, but of course I may feel differently if I get there.

Predicting the climate years ahead

Will global warming reach catastrophic levels this century?

Now we get to a period in the future that some of us will no longer be around to experience. The future of the climate has become a major societal issue, and predictions about what might happen under different policies have not only become politically important but have profoundly influenced the lives of many concerned individuals. Yet that future is intrinsically uncertain.

Climate models can be considered as generalizations of models for short-term prediction of the weather, as they are based on mathematical representations of how aspects of the world will develop according to physical laws. They not only consider movements of the atmosphere but also the dynamics of the oceans, the temperature of the land, and the ice on land and on the sea. These extraordinarily complex models necessarily have a much coarser representation than those for weather, both in terms of time and space, so that they can run through a century or more of change in a practical amount of time. They also differ in that they are not so sensitive to initial conditions, but will be sensitive to 'forcings' (external influences such as carbon emissions) and assumptions about how the climate will react. Just as with a weather model, ensembles can be created to represent possible futures, although the different members of the ensemble arise from perturbations of the important parameters of the model, rather than the initial conditions.

The Intergovernmental Panel on Climate Change (IPCC) 2019 Sixth Assessment Report (AR6) identified three major sources of uncertainty about the future climate:[20]

1. The natural and unavoidable variation in the climate. This can be approximated by running ensembles but is essentially irreducible.
2. The policies and actions taken by society. This is handled by modelling the consequences of a range of scenarios, from aggressive policies that bring net-zero carbon emissions by the middle of this century, to a 'business-as-usual' scenario in which emissions continue to increase.
3. How the climate responds to the actions of society. This is the trickiest part, as it involves multiple assumptions about the sensitivity of climate to changing emissions, and in particular any feedback in the system. This is often termed 'model uncertainty', although I have already said this is an inappropriate term, since no model could ever be correct. It could better be termed 'model indecision'.

The crucial third source is explored through looking at the results from over thirty different models constructed by teams around the world. Many of these models may arise from a common source and so have common biases, but in general there is no robust and reliable method for weighting models, and so the default is to use a single projection from each model and examine the variation between them. This is known as an 'ensemble of opportunity'.

As always, these models are imperfect, perhaps very imperfect, representations of reality. So for many projections, the IPCC takes the pragmatic approach of changing the interpretation of calculated uncertainty intervals; an apparent 'very likely' 90% interval (5% to 95%) derived from the spread of a multimodel ensemble is instead treated as a 'likely' 66% interval (17% to 83%) – this could be considered a further example of adding additional subjective 'Type B' uncertainty to allow for unmodelled factors.

An exception concerns the important metric of Global Surface Air Temperature (GSAT) where, instead of a rather arbitrary adjustment, additional uncertainty is numerically assessed using a wide range of sources, including the quality of agreements with historical observations. Some of these widened intervals are shown in Figure 11.3 for different scenarios.

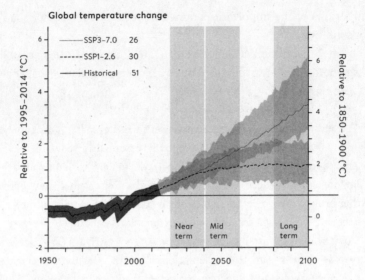

Figure 11.3

Estimated rise in average global surface air temperature by 2081–2100, relative to 1995–2014 and 1850–1900, under different emissions scenarios. The shaded bands show the calculated 'very likely' 5–95% ranges for scenario SSP1-2.6 (lower band – a moderate-emissions scenario achieving net zero in the second half of the century) and SSP3-7.0 (upper band – a high-emissions scenario). The numbers next to the scenario labels indicate the number of models used in the projections. Assessed by the IPCC in 2019.

A global warming level above 2°C since 1850–1900 levels was set by the Paris Climate Agreement in 2015 as a limit to avoid disastrous impact. Figure 11.3 suggests that even under the moderate emissions scenario SSP1-2.6, reaching net zero by around 2050, this threshold is still expected to be crossed by the end of the century, although with considerable uncertainty. However, this threshold is very likely to be exceeded under a high-emissions scenario (SSP3-7.0) by the 2040s.

A huge amount of work has gone into climate models, and their outputs can have a serious impact on societal concern and policy decisions. But any claims need to be accompanied by both numerical assessments of uncertainty and due humility about the models, even if sceptics try to take advantage of this uncertainty to criticize climate science. In the words of climate modeller David Stainforth, 'Even if the models can reproduce climate history, we shouldn't expect them to reliably tell us about the strange new future we're facing.'[21] Stainforth goes on to argue that building bigger and more complex models will not necessarily lead to greater insights, and may delude us into thinking we can predict what is going to happen, so 'instead of trying to make our responses just right for the climate of the future we should seek out resilient and flexible solutions, remedies that will be robust in a wide range of possible climate outcomes.' We shall return to this perspective when we come to deep uncertainty in Chapter 13.

————

We've covered a wide range of areas where predictions are routinely used, and it is time to take stock.

While simple methods may be adequate for short-term predictions in controlled situations, an industry has grown up creating increasingly complex models about the future. Many are used to try to make money: sports-betting companies predict outcomes using both past performance and in-game observations; hedge funds build sophisticated models for variability in the financial markets, hoping to profit from both rises and falls in value; while 'predictive analytics' are used to try to optimize business decisions. Models have different ways of handling uncertainty – some are fully stochastic, such as models of financial time series, while others are fundamentally deterministic and have the uncertainty 'bolted on', such as weather and some pandemic models.

The examples in this chapter demonstrate that uncertainty in prediction models comes from four basic sources:

1. Unavoidable aleatory variability, or chance, which cannot be reduced. This is sometimes termed 'random error'.
2. Epistemic uncertainty about the parameters of the model, both currently, and whether they may change in the future.
3. Uncertainty arising from the choice of structure of the model – as mentioned previously, this is not really uncertainty *about* the model, as there is no 'true' model that could ever be determined.
4. The systematic discrepancy of any model from the truth, over and above the unavoidable random error. This may be reducible by constructing a better model but can never be eliminated.

We've seen that when the future depends crucially on human behaviour, then we should feel far less confident in both the structure of the model and the appropriate assumptions about the parameters. Also, assumptions that are now reasonable about parameters and model structure may cease to hold in the future. Longer-term Covid-19 projections depended crucially on the timing of an effective vaccine, while financial models may have worked well during a stable period, when events were only weakly correlated, but failed disastrously when a looming financial crisis led to mass group behaviour.

Once we acknowledge the fragility of a specific model, we may be unwilling to derive confident probabilities for future events. So we are left with the approaches outlined previously: express caveats, conduct sensitivity analysis, engage multiple teams, use confidence ratings, and if necessary acknowledge deep uncertainty. It also makes sense to consider an ensemble of approaches, perhaps including prediction markets, expert subjective judgement, or even super-forecasters.[22]

Above all, express due humility about any claims, and be sceptical about anyone saying their analysis will tell you what is going to happen.

Summary

- In well-controlled circumstances, simple statistical models can provide reliable probabilities for future events.
- As we look further into the future, assumptions about the structure and the stability of models become more important.
- Uncertainty can be added to complex deterministic models by running ensembles, either from different initial states or with perturbed parameters.
- Even with every effort, models will be an inadequate representation of reality.
- Humility is needed, particularly when the future is strongly influenced by human behaviour or other unknowable factors.
- Insight can be gained by investigating a variety of prediction methods.

CHAPTER 12

Risk, Failure and Disaster

We start with a chilling example of what can happen in, literally, a 'perfect storm'.

Why did the MV *Derbyshire* sink?

MV *Derbyshire* was a huge bulk carrier of more than 90,000 tons, twice the size of the *Titanic*, and on 9 September 1980, during Typhoon Orchid off the coast of Japan, it just disappeared. No signals were received, no lifeboats were found, and forty-four people died in what was the largest British ship ever lost at sea. It was modern, built to current standards, and the reason for its loss was a mystery for two decades.

In 1994 a search finally revealed the wreck, 4km down and spread over more than a kilometre. Photographic evidence showed that the hatch covering the cargo hold at the front of the ship had suffered a catastrophic collapse, and the issue then became how sufficient pressure came on to the forward hatch to cause the failure.

Solving this mystery required some theory that started over a hundred years ago, when researchers in the British cotton industry realized that the risk of thread breaking depended on the strength of the weakest fibres. This meant that, rather than looking at *average* strength, statisticians needed to understand the variability in the *minimum* strength of a set of fibres in a thread. This requires special care. Standard statistical modelling tends to be concerned with explaining or predicting routine occurrences and so focuses on typical observations. But, as we saw in climate attribution studies, when we are interested in extremes the shape of the tails of a probability distribution becomes crucial. *Extreme value theory* was born.

In their contribution to the inquiry into the loss of MV *Derbyshire*, statisticians Janet Heffernan and Jonathan Tawn used extreme value theory to model potential pressure from waves, using data from experiments with a scale model of the ship and estimates of the size of waves in the typhoon. Specifically, they assumed the waves followed what is known as a **generalized Pareto distribution,** which, as we shall see below, permits previously unseen extreme events. They concluded that, if the ship had suffered some damage early in the typhoon, then it was very likely that an impact sufficient to collapse the front hatch would have occurred at some point.[1] A 'rogue' wave over 20m high may have hit the ship and caved in the forward hatch, followed by a rapid collapse of the other hatches and the ship sinking in seconds. The ship essentially imploded and then, as it sank, it exploded due to the compressed air between the hulls, scattering debris across a wide area. There was not even time to send a distress call.

After twenty years, the families of the victims finally knew what had happened.

The sad story of the MV *Derbyshire* could have been told in Chapter 10, since it is about trying to attribute causes to past events. But it is a suitable introduction into a discussion of extremes; earthquakes, floods, volcanos, terrorist attacks and major financial crises are all events with both low probability and high impact on society, and possibly of a type that may have never occurred before.

There are a lot of potential threats to think about, whether to society or to us as individuals. We might be concerned with environmental risks of climate change or extreme weather events, financial risks around the cost of living and pensions, health risks such as cancer or Covid-19, risks from modern technology such as AI, safety risks from malicious acts of violence or crime, and so on, and so on, right up to existential threats to the whole of humanity.

This may all seem a bit overwhelming and, unsurprisingly, a vast amount has been written about how to deal with the uncertainty surrounding all these perils, illustrating widely differing perspectives on the subject of *risk*. Approaches come under a number of broad headings, including

- *Technical*: **Quantitative risk analysis** (QRA) attempts to mathematically model both the probability and consequences of events in order to provide numerical input into decisions.

- *Economic*: According to basic economic tenets of rationality, risks can be handled by a theory of decision-making under uncertainty, which we shall come to in Chapter 15. However, this assumes a fully specified problem, and even then can be questioned.

- *Psychological*: We've already seen in Chapter 1 how perceptions of risks can depend on factors such as familiarity, dread, and so on, rather than the 'actual' probability of harm, and we know from our own experience that feelings about threats are complex and vary widely.

- *Cultural*: Groups of people can have very different views on how risks should be dealt with in society. In the Covid-19 pandemic, we saw some extremes of opinions, from a 'libertarian' perspective that individual behaviour should not be mandated by the state to a more 'hierarchical' view which encouraged strong interventions for the greater good. Risk becomes political.

- *Sociological*: In the post-Chernobyl 1990s the idea grew that a purely technocratic approach to risk assessment and risk management was inadequate, as it was dominated by professionals with limited insights and imposing their own worldviews. Since then the social context of risk has been increasingly acknowledged, whether through widening participation, accounting for justice and fairness, or acknowledging the interconnectedness between global threats.

My professional background as a statistician means that I focus on the more technical approaches to quantifying and communicating risks, but I recognize that this needs to be tempered with insights from psychology and sociology. These can help understand why some threats attract particular attention, and why people might be sceptical of a mathematical 'rational' approach to things that might inspire fear and loathing. However, it is unavoidable that discussion of extreme events can become rather technical.

———

We saw in Chapter 2 that super-forecasters may be reasonably good at assessing probabilities for plausible futures, but when it comes to the tiny probabilities of extreme events, we may expect pure subjective judgement to be poor. So statistical models are often used to try to quantify both the low probabilities and the high impacts, although these models can have all the limitations we've discussed previously. At the start of the financial crisis in 2007, the *Financial Times* reported David Viniar from Goldman Sachs as saying, 'We were seeing things that were 25-standard deviation events, several days in a row', which, assuming a normal distribution, are events with a probability of around one in 10^{135} (that's 135 zeros). To put this in perspective, the chance of winning the jackpot in the current UK 6/59 lottery is around 1 in 45 million, so an event with a probability of one in 10^{135} is like winning the jackpot seventeen times in a row. This strongly suggests the financial models were inadequate in their modelling of extremes.

The assessed probability for extreme events depends crucially on the shape of the tails of the distribution, and it turned out that the financial models were essentially assuming normal probability distributions, which have very 'thin' tails. But we have seen that statisticians exploit a wide range of distributions with 'fatter' tails than the normal curve, such as the generalized Pareto model used for the waves that hit the MV *Derbyshire*, whose tail has the form of a **power law** with a *shape parameter a*, meaning the probability distribution declines proportionally to $1/x^{a+1}$, where lower a corresponds to a fatter tail.*

Power-law tails occur in many situations when substantial variation includes some very large cases, such as the size of cities, the number of employees in a firm, stock market returns (with shape a around 3), while the distribution of the number of sexual partners in the previous year has shape around 2.5, indicating a long tail of activity.[2] Back in 1896, Vilfredo Pareto argued that the distribution of wealth followed a power law, after observing that 80% of the land in Italy was held by 20% of the people (a around 1.2), and economist Xavier Gabaix reports current shape parameters of around 1.5 for wealth and between 1.5 and 3 for income. Analysis of over 13,000 terrorism incidents between 1968 and 2007 found that the number of casualties followed a power law with $a \approx 2.4$ (95% interval 2.3 to 2.5),[3] with a very long tail reflecting occasional mass fatalities such as those following the 9/11 attack in New York. This would mean an attack such as 9/11, which caused over 2,700 deaths, would not be a particular outlier, with an assessed probability between 11% and 35% of occurring in this period. However, the estimate of the shape of the terrorism curve was presumably influenced by the 9/11 event, and so there is a degree of circularity in this calculation.

---

* Note that x might be a linear transformation of the actual measurement. A Student's t distribution with a degrees of freedom has tails of approximately this form.

In the previous chapter we saw that in May 2018 the Bank of England was projecting that the growth in the first quarter of 2020 would be around 1 to 2% per year, although the lower end of the 'fan' showed they assigned a 0.05 probability to a fall of greater than 1%. The actual fall, on an annual basis, turned out to be 25%.*

We emphasized that the Bank of England explicitly avoided modelling the extremes of their distribution, but it is informative to look at the consequences had they made specific assumptions about the shape of the tail. Figure 12.1 shows four possible options, with generalized Pareto distributions with shape parameters a = 1, 2 and 3, compared to a normal distribution.

Figure 12.1 (a) shows the distributions for the change in GDP, for falls greater than 1%. The area under each of the curves is 0.05, and they do not look particularly different to the naked eye. However, a somewhat different picture emerges if we look at the 'exceedance probabilities' in Figure 12.1(b) – these are the probabilities that the fall is greater than the value on the x-axis.† Each of these starts at 0.05, since all the curves are modelling the most extreme 5% of the distribution. But then the normal distribution drops rapidly, essentially ruling out extreme falls. Whereas the fatter-tailed Pareto distributions retain a reasonable probability of a really extreme event, with the curve with shape parameter 1 giving 0.006 probability to falls even greater than the 25% fall observed. Another interesting feature of the Pareto distribution is that, given you know the value has exceeded a certain 'failure' threshold, its conditional distribution is still a Pareto distribution with the same shape – if we condition on a fall of more than 1%, the curve with shape parameter 1 would assign 0.12 probability that the fall is greater than the observed 25% – in other words, if we are going to face an extreme event, the Covid experience was not particularly surprising.

———————

Once we have a method for assessing low probabilities for extreme high-impact events, it is natural to start making comparisons between all the risks we may face. In the early 1970s, amid growing concern about the safety of

———————

* Coincidentally, if we fitted a normal distribution to the central 90% of the fan chart, this would have been around 25 standard deviations from their 2018 central projection.

† If the probability distribution has a power-law tail proportional to $1/x^{a+1}$, then the exceedance curve also has the form of a power law, but proportional to $1/x^a$.

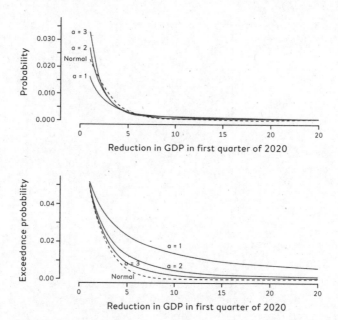

Figure 12.1

(a) The conditional distribution of the reduction in GDP in Q1 of 2020, given the fall is greater than 1%, assuming the distribution of tails form a generalized Pareto distribution with shape parameter a = 1, 2 or 3. These are compared to a normal distribution. All the curves contain 0.05 probability. (b) The exceedance probabilities, meaning the probability of observing a larger reduction.

nuclear power plants, the US Atomic Energy Commission hired MIT Professor Norman Rasmussen to examine the risk of accidents. Rather than just consider possible types of failures and their consequences, Rasmussen's team assessed probabilities for each step of the chain of events that could lead to failure, and so pioneered the use of probabilistic risk assessment[4] and its use in comparing threats.

Rasmussen's report was published in 1975 and prominently featured a type of diagram shown in Figure 12.2. These had been introduced in the 1960s by Frank Farmer of the UK Atomic Energy Authority in the context of the siting of nuclear power plants – apparently Rasmussen had discussed risk assessment with Farmer over 'viciously combative games of Ping-Pong at the British regulator's home'.[5]

The Farmer diagram plots the number of lives lost in an accident along the horizontal axis, on a logarithmic scale going upwards in multiples of 10, from

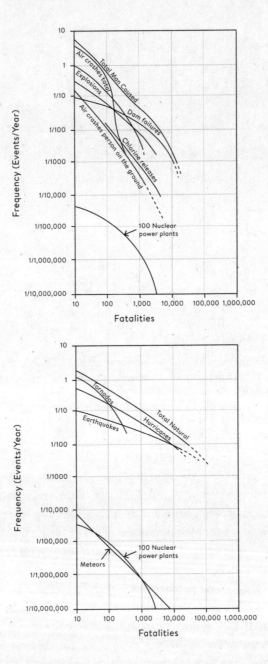

Figure 12.2
F-N curves (Farmer diagrams) from the 1975 Rasmussen report on the safety of nuclear power plants in the US, plotting the cumulative frequency of events against their severity in terms of fatalities, both measured on a multiplicative scale. The curve for 100 nuclear power plants was assessed to be substantially lower than for other man-made and natural hazards.

10 to 100 to 1,000, and so on. The assessed annual frequency (events per year) of such a severe accident is plotted on the vertical axis, again on a logarithmic (multiplicative) scale – from 1 in 10,000,000 years up to 10 times a year. Since they plot the cumulative Frequency of accidents against the Number of deaths, they have become known as **F-N curves**.[6] These are essentially exceedance probability curves plotted on multiplicative axes, and the area under the curve is the expected number of deaths. One consequence of using these axes is that power curves for the exceedance probabilities, as shown in Figure 12.1(b), conveniently turn into straight lines on an F-N diagram.[*]

Figure 12.2 shows around $\frac{1}{1,000,000}$ annual probability of a nuclear plant accident leading to more than 1,000 fatalities, compared to around $\frac{1}{100}$ probability for many other sources. This enabled Rasmussen's report to make some bold claims about the safety of reactors, saying 'that non-nuclear events are about 10,000 times more likely to produce large numbers of fatalities than nuclear plants'.[7] He also used Monte Carlo methods to assess the uncertainty arising from the many probability judgements underlying the assessments, reporting an uncertainty range of $\frac{1}{5}$ to 5 on probabilities and $\frac{1}{4}$ to 4 on consequences, so a claim of a $\frac{1}{1,000,000}$ annual risk of at least 1,000 casualties should be interpreted as between a $\frac{1}{200,000}$ and $\frac{1}{5,000,000}$ probability of at least 250 to 4,000 casualties.

Rasmussen's conclusions faced sustained criticism from the Union of Concerned Scientists and others, who accused the report of painting an over-optimistic picture and not properly allowing for uncertainty in the risk estimates. Even the Nuclear Regulatory Commission withdrew its endorsement of the main findings in January 1979. Then just two months later, on 28 March 1979, the reactor on Three Mile Island suffered a partial meltdown and 140,000 people were evacuated. Perhaps surprisingly, this accident tended to *increase* acceptance for Rasmussen's apparently optimistic report, since he had explicitly warned of the risks of local coolant loss and subsequent human error in the response, but had said that the health effects of such an accident would be expected to be negligible. Which is just what happened at Three Mile Island.

In retrospect, Rasmussen may have under-estimated the risks by insufficiently taking into account poor design, inadequate regulation, and human

---

[*] If F_n is the probability of getting at least n fatalities, then a power law means that $F_n \propto \frac{1}{x^a} = \frac{k}{x^a}$ for some k. Taking logarithms of each side gives $\log F_n = \log k - a \log x$. So on an F-N diagram plotting $\log F_n$ against $\log x$, a power law will be a straight line with gradient $-a$.

factors such as employee overwork and lack of training – all issues that con-tributed to subsequent crises at nuclear plants around the world. But he put probabilistic risk analysis on the map and this led to a surge in models for assessing the risks of everything from floods to food poisoning.

———————

Insurance companies will only be profitable if they have a sound judgement about the frequency and cost of accidents, and so they develop models for well-understood areas such as car, life and travel insurance, using extensive databases of people who have crashed, died and fallen down steps on holiday. The potential financial costs are (relatively) not great.

It's more challenging when it comes to disasters. When Hurricane Andrew hit Florida in 1992 it caused nearly $30 billion in damage, and the huge losses made by insurance companies showed a more sophisticated mod-elling approach was needed, both to assess premiums to customers and to negotiate 'reinsurance', in which insurance companies transfer the risk by further insuring themselves against large losses. This led to the development of *catastrophe modelling* (often shortened as cat-modelling), in which a proba-bilistic model may be used to simulate a large database of potential disasters, each with associated impacts and costs.

The main output is then an exceedance probability curve showing the probability of losses being above each of a range of possibilities. But there are the usual sources of uncertainty: doubts about the underlying physical mechanisms, expert disagreement on assumptions, poor data, unavoidable limitations of any model, and so on. As usual, responses may include verbal caveats, sensitivity analysis (for example showing multiple exceedance prob-ability curves), adding subjective judgement into the analysis, and multiple independent models, although this seems to be one area where scales of 'con-fidence' have not been adopted.

When it comes to floods, extreme value theory can be used to estimate the annual risk of a certain level – if this is estimated to be, say, 0.1%, this is often translated as a '1 in 1,000-year event', also known as a *return period*, and defences can be built to protect against these levels. This is the standard of protection applied to many flood defences in the Netherlands, although near cities a more stringent '1 in 10,000 years' (0.01% per year) is demanded.[8] There are many uncertainties in these assessments, both because of the sensitivity to the model and the fact that the environment is changing; a recent report estimated that in thirty years' time the probability of flooding would be ten times higher in more than a quarter of the places studied.[9] These uncertainties can lead to precautionary measures such as an additional margin of safety on heights of barriers.

While 'return period' may be the accepted technical way of describing a small annual risk, it can be a misleading public communication, as after a '1 in 100-year' event has happened, people may assume the next one is not going to happen for another hundred years, and they complain when it happens again after just a few years. I live metres from the River Cam, and the flood-risk map[10] tells me my house is in a 'low risk' area for river flooding, which used to be explained as between a '1 in 1,000-year' and '1 in 100-year' risk, but the UK Environment Agency has now stopped using return periods in their communication, and the risk to my area is now translated as 'between 0.1% and 1% per year'. I am fairly relieved to hear this. I think.

————

You may have had to make risk assessments at work or when planning events, and managers in many organizations have to consider a list of potential threats and assess their 'likelihood' and impact. The UK is no exception, and the UK National Risk Register* assesses the most serious short-term acute risks facing the UK or its overseas interests – chronic long-term threats such as climate change and artificial intelligence are dealt with separately. The 2023 register[11] was considerably more detailed and transparent than previous versions, and for the first time assigned numerical likelihoods for malicious risks such as 'strategic hostage taking' (given probability 0.2–1% over 2 years).

Risks are evaluated as 'reasonable worst-case scenarios' (RWCS), defined as 'the worst plausible manifestation of that particular risk (once highly unlikely variations have been discounted)'. According to the UK government yardstick, 'highly unlikely' means 10–20%, so this could mean that the RWCS corresponds to around the 80–90% point of severity, with only around 10–20% probability of something worse happening. But in practice the term seems to be used informally, with the idea that contingency planning should be based on pessimistic, but still plausible projections – a concrete manifestation of the old saying 'hope for the best, plan for the worst'.

---

* This is the public-facing version of the (extremely non-public) National Security Risk Assessment.

Impact					
5. Catastrophic	Civil nuclear accident		Failure of national electricity system	Pandemic	
4. Significant	Aviation collision	Reservoir/ dam collapse		Outbreak of an emerging infectious disease Severe space weather	
3. Moderate	Large passenger vessel accident			Volcanic eruption	
2. Limited		Strategic hostage taking	Public disorder		Assassination of a major public figure
1. Minor	Earth-quake				
	1: < 0.2%	2: 0.2–1%	3: 1–5%	4: 5–25%	5: > 25%

Likelihood

Figure 12.3

A sample of entries in the 2023 UK National Risk Register. 'Likelihood' is the assessed probability of the reasonable worst-case scenario occurring at least once in the assessment period (2 years for malicious risks; 5 years for non-malicious risks). 'Severe space weather' occurs when a burst of charged particles from the sun is carried by the solar wind into the Earth's magnetic field, and can lead to satellite damage, radio outages and power failures; the solar storm in 1859 paralysed the telegraphic system, giving some operators electric shocks.

Figure 12.3 reproduces part of its main 'risk matrix', in which the RWCSs of various threats are assigned categories of impact and likelihood, with impact being defined on a five-point scale; for example a RWCS would be given an impact of 3 if it led to 41–200 deaths, 81–400 casualties or a cost of hundreds of millions of pounds. The likelihood scale increases five-fold at each step, and this is reflected in the increasing width of each box in the graphic – this has been shown to increase comprehension of the non-linear scale.[12]

Risks are placed in fairly broad bands, but there are two reasons why even this may be too precise. First, assessing the impact and likelihood of a reasonable worst-case scenario is challenging, and the register allows uncertainty about its position on the matrix, accounting for limitations in the evidence, reliability of assumptions, and external factors that may influence what happens. For example, a severe space weather event would be expected to cause significant impact, with substantial 5–25% probability over five years, but could extend to adjacent categories on each scale.

Second, we may want to get an idea of the spread of possibilities rather than just focus on the extreme RWCS – we have already seen how this led to problems during the Covid-19 pandemic, with modellers being accused of being overly pessimistic, after they had been explicitly tasked with modelling such scenarios. An even worse example was in the swine flu epidemic of 2009, when a RWCS of 65,000 deaths was announced by the Chief Medical Officer, and prominently reported as 'Swine flu could kill 65,000'. It later turned out that this was based on an outdated and very pessimistic analysis, and neither the modellers nor the Chief Medical Officer thought this figure was 'reasonable'. The final death toll was 460.

A subsequent parliamentary inquiry concluded the government 'should use the experience of the 2009 pandemic to emphasize the range and likelihood of various possibilities',[13] although this lesson does not appear to have been learned. Perhaps the risks should be represented as 'blobs' on the risk matrix, representing the variability of possible future events.

The selected sample of threats can be rather sobering. A pandemic is still considered the most likely catastrophic event, while assassination of a major public figure is assigned more than 25% probability over the next two years, but with a minor impact. Fortunately an earthquake is not considered a serious peril – in the UK – and you may think the threat to the UK from volcanos would be rather negligible, given there aren't any active volcanos anywhere near the British Isles. But volcanos can still be a threat . . .

What is a reasonable probability of an effusive volcanic eruption in Iceland occurring and having a major impact on the UK?

In spring 2010 the Icelandic volcano Eyjafjallajökull erupted, throwing out huge quantities of volcanic ash which was carried south over Europe, halting air travel, stranding 10 million passengers and causing massive disruption. Volcanos did not feature in the 2008 Risk Register and no plan was in place for the UK, and only very rapid negotiations between engine manufacturers, airlines and regulators enabled flying to the UK to resume.

There was substantial economic damage from Eyjafjallajökull's explosive eruption, but that was trivial compared to the harm that could be caused by a more gentle but continuing 'effusive' eruption, producing vast lava flows rather than explosions. This occurred in Iceland in 1783–4, when the Laki fissure poured out sulphur dioxide, chlorine and fluorine over eight months. The poisonous cloud killed half the livestock on Iceland and led to the deaths of a quarter of the population. It then darkened the skies and rained sulphuric and other acids over Northern Europe, leading to crop failure, widespread hunger and the deaths of thousands. It is said to have helped precipitate the French Revolution in 1789. So the impact was a lot more serious than inconveniencing air passengers.

In the wake of the 2010 eruption, I was part of a team asked to help assess a reasonable worst-case scenario for an effusive eruption in Iceland. This required taking a very long-term perspective, as geological evidence pointed to previous less severe eruptions over a thousand years ago in 934, and again in 1612, suggesting a return period of hundreds of years. Effusive volcanos subsequently appeared in the 2012 Risk Register as having Impact 4 and a likelihood of between 1 in 200 and 1 in 20 in the next five years (using the 2012 labels), i.e. between 0.1% and 1% in each year. In the 2023 Register summarized in Figure 12.3, volcanic eruption is considered as a single scenario with Impact 3 and a likelihood between 5% and 25% over five years, so a Laki-type catastrophe is not considered separately. Let's hope it doesn't happen.

Having lived through the Covid-19 pandemic starting in 2020, it is informative to look back and see what was in the 2017 Risk Register that covered this period. While a flu pandemic was given the highest possible impact, an 'emerging infectious disease' such as SARS and MERS was only given Impact 3, with 'several thousand people experiencing symptoms, potentially leading to up to 100 fatalities'. This was the underestimate of the century – by the end of 2023, over 230,000 people in the UK had died with 'Covid-19' on their death certificate. Perhaps a greater acknowledgement of uncertainty and variability in 2017 would have been useful.

_____

The UK Risk Register looks at most five years ahead, and uncertainty only grows as we consider the longer-term future for a country or the world. The World Economic Forum conducts an annual survey of what respondents feel are 'global risks',[14] and its 2023 league table, using a ten-year horizon, prominently features natural disasters, climate change, involuntary migration, erosion of social cohesion, cybercrime and 'geoeconomic confrontation'. They make no attempt to break these risks into likelihood and impact and so these ratings simply represent concern, but it is notable how these issues tend to be complex and interconnected – so-called 'wicked' problems, which are not readily defined and have no straightforward solution.[15]

Although these judgements may serve to raise awareness of concerns, it is difficult to feel much confidence in their magnitudes. Which brings us nicely to the question that you may have been increasingly pondering throughout this chapter – can we really believe all these risk analyses?

_____

Ever since the innovative work of Rasmussen on the threats from nuclear power plants, there has been vociferous criticism of quantitative risk analysis, largely based on the following running themes.

Spurious precision and unknown accuracy. When there were complaints about a proposal to build a cable car across the Thames beneath the flightpath to London City Airport, the National Air Traffic Service estimated an accident risk of 'approximately' 1 in 15,397,000 years. But the chance of an error in the assumptions, or even in computation, will dwarf this, and so this degree of precision may give an unwarranted impression of accuracy.* While we can routinely check the probabilities given to tomorrow's football results or next week's weather, we can't check the tiny numbers given to low-probability, high-impact events.

Limited scope. Quantitative risk analysis does what it says on the tin – it

_____

* These concerns have not prevented me from enjoying the cable-car trip.

puts numbers on both the probabilities of events and the severity of outcomes. But there are many aspects of the outcomes that are not so quantifiable. Philosopher Jonathan Wolff worked with the UK Rail Safety and Standards Board to help them understand why there was such pressure to further reduce risks when rail travel is extraordinarily safe, with just nine passengers killed in train accidents out of 23 billion journeys between 2005 and 2020. Wolff concluded that simply counting fatalities ignores important societal attitudes about the *cause* of the fatalities; deaths that are imposed on 'innocent' people, especially if due to failures in safety procedures, increase feelings of shame and outrage that such events should happen, and an urge to blame those seen to be responsible.[16]

Uncertainties. Just like in any modelling process, quantitative risk analysis should include unavoidable aleatory uncertainty, epistemic uncertainty about parameters and basic scientific understanding, and acknowledged limitations in the whole ability of any model to represent reality.

Personally, I believe that quantitative risk analysis can play a vital role, provided that the issues listed above are recognized. But it both focuses on well-characterized problems and rests on numerous unverifiable assumptions. So it is important to consider what to do when we admit that our very conceptualization may be inadequate, and when we do not feel happy about listing the possible futures, let alone putting probabilities on them. As we shall see in the next chapter, this leads us naturally into the mysterious territory of deep uncertainty.

Summary

- Risk is a complex area, with many different professional perspectives.
- Quantitative risk analysis attempts to place probabilities on extreme events and assess their impact.
- Fat-tailed distributions are important for modelling potential extreme events such as 'perfect storms'.
- Exceedance probability curves are used when modelling potential catastrophes for insurers, and F-N curves provide a way of comparing extreme risks.
- Risk matrices can display the rough likelihoods and impacts of a wide range of potential threats, but focusing on a 'reasonable worst-case scenario' neglects variability in potential futures.
- Quantitative risk analyses have been criticized for their spurious precision, their limited scope and their insufficient acknowledgement of uncertainty, and these concerns should be taken seriously.

CHAPTER 13

Deep Uncertainty

'There are more things in heaven and earth, Horatio, than are dreamt of
in our philosophy.'

—William Shakespeare, *Hamlet* (First Folio)

History is replete with examples where confident claims or decisions have
been contradicted by reality. Thomas Malthus notoriously predicted in 1798
that increasing population would inevitably lead to famine, but didn't take
into account the massive boost in productivity from the agricultural and
industrial revolutions. Two centuries later, his error was echoed by Paul Ehr-
lich's famous book *The Population Bomb*, published in 1968, that began by
saying that in the 1970s hundreds of millions of people would starve to death.
Needless to say, this did not happen.

The MV *Derbyshire* sank because of unforeseen pressure from rogue waves,
while the sea wall at the Fukushima nuclear power plant was designed to deal
with a wave height of 5.5 metres, not the 15-metre tsunami that hit it on 14
March 2011. Less well known is the administrative disaster of the 2012 'renewable heat incentive' in Northern Ireland, which would pay people to install
heating using renewable energy, at a projected cost of £25 million. Savvy
people realized they could apply to heat properties such as garden sheds that
had not been previously heated, and the 'cash for ash' scheme, as it became
known, had run through £500 million before it was halted. The scandal led to
the collapse of the Northern Ireland devolved government, which was not re-
established until 2024.

The originators of all these plans and ideas would presumably have been surprised at what actually came to pass. We can identify two broad sorts of surprises. First, *perfect storms* are an extreme version of a familiar event in the far tails of a distribution, such as the wave that hit the MV *Derbyshire*. Second, *black swans*, a term popularized by Nassim Taleb,[1] are qualitatively different types of events that had not even been thought of. For the kind of risk register we saw in the last chapter, perfect storms represent an extreme example of something that featured on the register but had not been assigned sufficient impact, such as 'emerging infectious diseases' only being assigned impact level 3 in 2017. Black swans are not even on the register, such as the Icelandic volcano in 2010.

If we want to be ready for surprises, we need to think beyond the reasonably well-understood areas examined in previous chapters, where we can list the possibilities and attempt to put our uncertainty into numbers. When we are facing circumstances which we don't understand well, we may be unwilling to assign probabilities, and have low confidence in any models. We might be considering events that have not happened before, when we might not even be able to imagine what shocks may be in store – just think of trying to specify the possible impacts of AI in twenty-five years' time. This has become known as deep uncertainty.

Of course, this assumes people have the insight and humility to admit when they don't know, whereas we've seen in the stories above that they may be deluded, and either explicitly or implicitly assume they have thought of everything and understand what is going on. This arrogance has been termed *meta-ignorance*, when people don't know that they don't know – what Donald Rumsfeld identified as 'unknown unknowns' – and should lead to serious shock when something surprising occurs.

It is slightly less culpable to acknowledge uncertainty, but smugly assume you have got it nicely quantified and modelled. When something unexpected occurs, then you just elaborate the models and assume that you now have it sorted – until a surprise happens again. As we've repeatedly emphasized in this book, it would be better to acknowledge that models are imperfect but to specify areas of inadequacy and learn from experience.

There is no 'correct' level of humility – we need to choose what is appropriate for each situation. If we have too much confidence, then we risk being caught out by events, and perhaps examining our deeper uncertainties would be more valuable than more data collection and analysis.[2] But if we overstate our ignorance we may lose the valuable insights of quantified models – just because we don't know everything, it does not mean we don't know anything.

In particular, under-estimating our knowledge and understanding may lead to excessive caution, a topic we shall return to in Chapter 15.

If we seriously want to acknowledge our ignorance, then we can own up to what is sometimes termed **ontological uncertainty**. This is when we admit that our whole conceptualization may be inadequate – that the potential outcomes, important features, underlying ideas, assumptions and the very language being used are all questionable. This may seem a major step, but is a natural consequence of admitting that we do not experience reality directly, and only through our senses – we then try to structure these experiences through concepts, thought and language, which we finally try to communicate to others. All concepts are essentially models – the map, not the territory – which we should know by now do not represent reality and are all 'wrong'.

So we really should have the humility to admit ontological uncertainty and, for example, emphasize the 10% that is not contained in the main part of a Bank of England fan chart. Although this is easier said than done – we can read Hamlet's line at the start of this chapter, but it is difficult to think outside our own personal box. Which is why it is so important to have diversity of viewpoints, so our pre-set ideas can be brought under scrutiny.

> What do we do when we are not happy to list potential outcomes and assess their probabilities?

There is no sudden jump into deep uncertainty; we wade in along a continuum in which there are increasing difficulties in both specifying possible outcomes and assessing probabilities through judgement and models. Figure 13.1 provides a simplified view of this process, indicating possible approaches we might take within each of the four quadrants, which we shall consider in an anticlockwise order.

Quadrant (A) represents the standard 'reductionist' approach in which we model both what could happen and assign probabilities to potential events, of course with appropriate precision, and acknowledging limitations in the analysis and resulting uncertainty, just as for the threats in the National Risk Register.

Quadrant (B) includes situations where people are unwilling to put uncertainty into numbers, even though it may be about well-defined possibilities. Economists frequently quote two authorities from the first half of the last

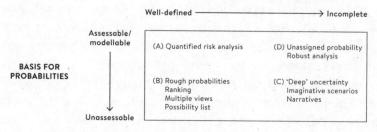

Figure 13.1
Approaches to dealing with risk and uncertainty as we become increasingly unsure about our assumptions and assessments. Based on a proposal by Andy Stirling.[3]

century in support of this position, but I strongly disagree with both of them. First, John Maynard Keynes wrote in 1937:

> By 'uncertain' knowledge, let me explain, I do not mean merely to distinguish what is known for certain from what is only probable. The game of roulette is not subject, in this sense, to uncertainty; nor is the prospect of a Victory bond being drawn. Or, again, the expectation of life is only slightly uncertain. Even the weather is only moderately uncertain. The sense in which I am using the term is that in which the prospect of a European war is uncertain, or the price of copper and the rate of interest twenty years hence, or the obsolescence of a new invention, or the position of private wealth-owners in the social system in 1970. About these matters there is no scientific basis on which to form any calculable probability whatever. We simply do not know.[4]

I would argue that these anachronistic claims were rendered out-of-date when ideas of subjective probability became both respectable and widespread. Keynes would presumably acknowledge that in 1957 copper would have a price, and there would be a rate of interest, assuming the continuation of capitalism, so these are well-defined and eventually observable quantities. He says there is no basis for *calculating* a probability, but that does not mean there is no way of *assessing* a probability – just the sort of task to set to a super-forecaster who could take advantage of long historical series.

Second, in his classic 1921 book *Risk and Profit*, economist Frank Knight considered 'risks' as objective quantities that could either be obtained by reasoning (for example using the symmetries of dice or cards) or estimated from historical data, and claimed, 'It will appear that a measurable uncertainty, or "risk" proper, as we shall use the term, is so far different from an unmeasurable one that it is not in effect an uncertainty at all', and that uncertainty, in contrast, concerns partial knowledge for which 'the conception of an objectively measurable probability or chance is simply inapplicable'.[5] Again, Knight is only focusing on situations where there is no 'measurable' probability, and ignores the use of subjective judgements. The unfortunate phrase 'Knightian uncertainty' has come to be used for situations when people 'don't know the probability distribution', but this inappropriately implies that probability is an objective property of the world which we happen not to know.[6]

So in quadrant (B), where the potential outcomes are well defined, I believe there is almost always an opportunity to elicit judgements from skilled probability assessors. Of course, in some situations people may express such low confidence in their judgements that other methods can be adopted. These might include:

- Allowing ranges for probabilities: For example, as explored in detail in Chapter 2, terms such as 'likely' may correspond to a probability between 55% and 75%.
- 'Uncertainty tables': When assessing the uncertainty about scientific claims, the European Food Safety Authority's scientific panel are asked to be as quantitative as possible, using subjective judgements if appropriate.[7] But if experts find it impossible to quantify the influence of factors on their final conclusions, they can use symbols ↑, ↑↑, ↑↑↑ to represent minor, intermediate and strong upward influence on probability.
- A simple ranking of possibilities in terms of their perceived likelihood, without explicitly assigning probabilities.
- Just listing possibilities.

There have also been numerous attempts to develop whole new formalisms for 'unknown probabilities', such as imprecise probabilities specified as an interval, 'belief functions' or 'possibility calculus',[8] but these have had little take-up, possibly because of their complexity and the resulting wide range of conclusions.

———

Quadrant (C) is the trickiest – when risk assessors admit they are neither happy with specifying the potential outcomes nor expressing uncertainty in numbers.

This is quite reasonably described as deep uncertainty – situations that economists John Kay and Mervyn King[9] term *mysteries*, full of vagueness and indeterminacy and potential surprises, rather than the *puzzles* that can be handled using standard quantitative methods.*

Many efforts are made to avoid being taken by surprise by events – an exercise sometimes known as 'de-blackening'.[10] It sounds straightforward – just imagine all the horrible things that could happen, and plan for them, and then you can't be taken by surprise. Organizations have developed numerous techniques for scenario planning; for example, 'intuitive logic' follows causal chains forwards, while 'backwards logic' goes the other way, by starting from an undesired endpoint and considering all the ways that it could come to pass.[11]

But this all requires imagination and an open mind, and raises two crucial issues. First, that the scenarios should include careful consideration of people's 'reflexive' responses to a new situation, just like the heating-policy shambles in Northern Ireland was driven by public self-interest, and social behaviour altered the course of the Covid-19 pandemic. Scenarios should consider all aspects of a possible future, particularly who is affected and how they might respond.

Second, that diverse sources should be used to generate a wide range of possible futures, so that different viewpoints can challenge assumptions, identify blind spots and prevent group think in an organization. A standard strategy is to bring in a 'red team' to take an aggressive and pessimistic view, as they did when assessing the chance that Osama bin Laden was in the Abbottabad compound. Even more effort may be needed to reach people who may be able to imagine a potential black swan.

The IPCC has explored climate change scenarios representing ways that society might develop up to 2100, for example the 'B1 scenario' imagines a world with technological solutions to economic, social and environmental sustainability, which improves equity but does not have additional climate initiatives.[12] The scenarios are not assigned probabilities. Such scenarios can also feel a bit dry, and if they can be turned into more vivid narratives then people might engage emotionally with the stories, and this could lead to further insights. An interesting innovation is the UK Ministry of Defence (MoD) sponsoring science-fiction

---

* Kay and King prefer the term *radical uncertainty*. They distinguish this from *resolvable uncertainty*, which can be handled using stable (or 'stationary') probability distributions, whereas they claim radical uncertainty arises from distributions that are non-stationary over time and so cannot be adequately understood and modelled.

writers to produce a series of 'Stories from Tomorrow' about possible future events.[13] One of these (rather good) short stories, 'Silent Skies', comprises mock news reports from 2040 of a mass drone attack on London which was concealed in the huge volume of commercial drone traffic, leading to the commercially run 'Metropolitan Airspace Management' being taken over by the Ministry of Defence.*

We all tell ourselves stories, and Kay and King suggest that we have a reference narrative that summarizes how we would like, and expect, things to develop in the future. A 'risk' is anything that threatens to disturb this personal story. This sounds plausible, and it is an interesting exercise to explore our own reference narrative, which, if you are like me, is generally unexamined. Of course, such an overarching narrative can become too dominant and prevent flexible thinking – remember the foxes and hedgehogs in Chapter 2.

Kay and King argue that it is more useful to identify important risks that could disrupt our reference narrative, rather than expending effort trying to assess detailed probabilities. However, even in Quadrant (C), deep uncertainty does not eliminate the value of models, provided they are not taken too seriously as representing reality – they can be used to explore important individual aspects of the possible futures, and the possible effects of interventions. Exploring a full range of scenarios may be more important than modelling any of them very well. This reflects the experience during the Covid-19 pandemic, where unpredictable human behaviour rendered detailed predictions fairly pointless, but models provided a basis for judging broad potential impacts of measures against the virus.

The final quadrant (D) is curious, as it represents situations where there are problems specifying the potential outcomes, and yet a willingness to assign probabilities. This may seem contradictory but arises from a simple trick: lump together everything that is not listed, call it 'something else', and give it a probability!

We have already seen the Bank of England do this in their fan charts showing their uncertainty about future growth (see Chapter 11), where the lower tail contains 5% unassigned probability, and 'something outside our model'

---

* The stories carry the reasonable disclaimer 'The events, statements, and views expressed are similarly fictitious and should not be taken as a definitive or indicative view of, nor an endorsement by the Ministry of Defence.'

includes the sort of precipitous drop seen during the Covid-19 pandemic. And we showed in Chapter 7 that within a Bayesian framework, Cromwell's rule means that if we give a small probability to 'something else', this leads to an automated learning process in which we naturally adjust our beliefs after observing apparently surprising events.

Perhaps we should always include 'none of the above' in any list of possibilities, and be prepared to assign it a probability.

––––––––

Once organizations recognize they are facing deep uncertainty, they need to be ready for the unexpected and try to be resilient to whatever might come their way. At the same time, they need to be able to take advantage of new and unforeseen experiences, and not be so cautious that they become too anxious to act.

These ideas seem to apply to individuals as well. This is not a self-help book, but acknowledging our deep uncertainty and keeping an open mind might be helpful in our personal lives.

Summary

- In order to avoid being taken by surprise, we have to admit deeper uncertainty.
- We may have problems specifying possible outcomes, as well as assigning probabilities to events.
- It does not make sense to say 'we don't know the probability' – all probabilities are constructions, and we may be able to make subjective judgements.
- It is useful to construct scenarios, but this requires a diversity of imaginative viewpoints.
- Fully formed narrative stories may engage people, but fixation on a particular reference narrative can lead to 'hedgehog-like' unwillingness to adapt.
- By assigning a probability to 'everything else', we can formally handle situations in which we cannot pre-specify all eventualities.

CHAPTER 14

Communicating Uncertainty and Risk

In the aftermath of the tragic events of 11 September 2001, there was increasing rhetoric about the dangers posed by the Iraqi regime. In August 2002 US Vice President Dick Cheney said to the Veterans of Foreign Wars national convention, 'Simply stated, there is no doubt that Saddam Hussein now has weapons of mass destruction',[1] while reports from both the US and UK governments provided a 'case for war' with confident claims that Iraq had programmes to develop such weapons, including nuclear ambitions.[2]

When the US-led coalition invaded Iraq in March 2003, these claims were found to be false – neither active weapons of mass destruction nor efforts to restart a nuclear programme were found. A 2004 UK review concluded that expressions of uncertainty, present in the original non-public intelligence assessments, were removed or not made clear enough in the public version.[3] A US Senate Select Committee investigation went even further in its criticism of the wording of the intelligence about Iraq's capabilities, concluding that 'The Intelligence Community did not accurately or adequately explain to policymakers the uncertainties behind the judgments in the October 2002 National Intelligence Estimate.'[4] The omission of expressions of uncertainty from both documents may, through increasing the apparent threat posed by Iraq, have had a major effect on public opinion and government action in the lead-up to the war.

There are also plenty of cases where, in contrast, uncertainty has been exaggerated. From the 1950s, as evidence about the harms of smoking accumulated, tobacco companies developed a carefully formulated campaign to promote uncertainty and undermine trust in the science. The label 'Merchants of Doubt' has been applied to those who try to obfuscate scientific discussion about controversial topics,[5] although a slightly less catchy term for the deliberate cultivation of ignorance is 'agnotology'.[6] We saw all these tactics being used to cast doubts on much of the science of Covid-19, with social media contrarians deliberately cultivating uncertainty by saying they were 'just asking questions' about, say, large numbers of excess deaths linked to vaccines, even though those claims could be easily countered.

Consciously under- or overstating uncertainty could be considered as deliberate *dis*information. In contrast, we might consider it *mis*information when someone unknowingly overlooks appropriate uncertainty in their communication. As we saw in Chapter 13, people might simply be in a state of delusion or meta-ignorance – they genuinely don't know that they don't know. And it's not just people that can be unaware of their ignorance – as we shall come back to in Chapter 16, AI chatbots express opinions with extreme confidence, and yet can also 'hallucinate' claims that are manifestly untrue.

––––––––––

In contrast to the manipulations described above, let's assume that you are a communicator who wants to act with integrity and honesty, to help audiences understand what is going on and enable them to make decisions that fit with their aims and values. And even if we are not doing the communication, we are all on the receiving end of claims and need to decide whether we trust what we hear.

Trust is a crucial issue, and authorities all want to be trusted. I've been asked many times how organizations can increase and retain trust. But, as Kantian philosopher Onora O'Neill has emphasized, this is not an appropriate goal – rather they should be seeking to *demonstrate trustworthiness*.[7] Audiences may then choose to offer up their trust, and the organizations will deserve it. In short, their ethos is crucial to how they communicate, and being trustworthy should be the dominating theme.

But what to do? Psychologist Baruch Fischhoff makes clear that there is no 'correct' way to communicate risk and uncertainty – it depends on what you are trying to achieve. Once you have decided your aim, 'communicating uncertainty requires identifying the facts relevant to recipients' decisions, characterizing the relevant uncertainties, assessing their magnitude, drafting possible messages, and evaluating their success'.[8] This emphasizes that communication is not something to be left up to intuition, but requires systematic analysis, just as other human skills like listening and parenting can be improved by reflection and guidance. So this chapter will step through the whole process of communication, from the context, participants, aims and content to its effect on audiences, setting out a list of questions that any communicator should be able to answer.

Although there is no correct way to communicate, there are still some general principles for openness and honesty. Again, Onora O'Neill has come up with a useful list covering what she terms 'intelligent transparency', which requires information to be

- *accessible* to audiences – this should be reasonably straightforward in a digital age.
- *intelligible* to as many as possible, and this needs to be tested.
- *useable* to answer the concerns of audiences, which means careful listening.
- *assessable* by those who wish to check your work and have sufficient skills to do so.

The final point is easily overlooked, but vital. Most people may take your reasoning on trust, but if some specialists want to drill down, they should be able to reconstruct what you have done. This suggests 'layered' communication, in which further detail is provided for those who want it.

Trustworthy communication is profoundly important, and an integral part of that trustworthiness is conveying appropriate uncertainty about claims. This requires both humility and insight into one's aims and motives – so-called critical self-reflexivity. These characteristics will be celebrated as we work through the questions that need to be addressed.

What is the context of the communication?

Communication may take place in a range of situations, which might be *routine*, such as releasing economic data or drawing conclusions from scientific studies; *emotionally charged*, such as providing information to a patient newly diagnosed with cancer, or describing risks of carcinogens; or even in *crises*, such as when a disaster threatens, as in hurricane forecasting, or is actively in process, such as during a developing pandemic.

How should authorities communicate in a crisis?

Crises understandably receive a lot of attention and help to illustrate many important subsequent issues. We've already seen how public reaction to rail and other accidents has been characterized by both an urge to assign blame and a sense of shame that such events could happen in our society, and it is clearly essential for any communicator in a crisis to have real empathy with the feelings of the audience. Zoologist John Krebs faced a deluge of crises when he was chairman of the UK Food Standards Agency in the 2000s, including dioxins in milk and bovine spongiform encephalopathy (BSE, or 'mad cow disease'). He adopted the following five-stage strategy when communicating to the public,[9] telling them

1. what we know (knowledge);
2. what we don't know (uncertainty);
3. what we are doing to find out (plans);
4. what people can do in the meantime to be on the safe side (self-efficacy);
5. and that advice will change (flexibility and provisionality).

For example, at a press conference he admitted they did not know whether BSE had got into sheep, but said they were developing a diagnostic test, and in the meantime they were not advising people to stop eating lamb but, if worried, 'change your diet, and we shall get back to you'. There was no panic, and an initial drop in lamb consumption was counteracted when the price was reduced.

We are used to populist commentators expressing absolute confidence in who to blame for the latest crisis, so when I hear anyone talking to the public I check them against John Krebs's excellent list. My personal experience is that politicians find it extremely difficult to admit uncertainty (point 2), and are even more challenged by admitting the provisionality of advice (point 5) – they seem to feel they have to speak with absolute and unchanging conviction. These two issues are intimately connected – if there is no acknowledgement of uncertainty, then any change of policy opens them up to accusations of 'U-turn', and so binds them into decisions that may have become clearly inappropriate. Sadly we saw this during the Covid-19 pandemic, when people were still obsessively wiping surfaces long after it was clear that the major route of transmission was airborne rather than droplets.

Krebs's third point – what we are doing to learn more – would seem to be self-evident, but politicians generally appear reluctant to admit the need for research or even experiment. Again, in the Covid-19 pandemic, numerous policy decisions were made about dealing with infections in schools, with minimal supporting evidence. In the UK, a proper scientific study was finally carried out in 2021 of alternative strategies after a student tested positive for Covid-19, in which 201 schools were randomly allocated to either sending home all school contacts for isolation for ten days, or letting contacts stay at school if they tested negative each day for the following week.[10] It was found that both policies produced similar rates of infections in both students and staff, showing that vast numbers of students had unnecessarily been isolated after a contact became infected.

When the next crisis happens, check whether Krebs's list is being followed.

Who are the audience?

The first rule of communication is to shut up and listen. It is vital to understand your audience – their culture, their needs, their knowledge, their emotions and anxieties, their misunderstandings, their aims – whether they are your peers, the public, or politicians and decision-makers. The language and imagery must be appropriate – the Bank of England fan charts are good examples of where some effort has been paid to make the message accessible.

It's also important to realize that the audience's objects of uncertainty may be different from the communicators'. Organizations may produce estimates, intervals and expressions of confidence about a specific quantity, such as national GDP, but their audience may be more interested in how the economy is going in their particular area, or the impact of Brexit or other events. Again, this understanding only comes through listening to concerns.

It has become a cliché that one particular audience does not appreciate uncertainty – politicians. When presented with a range around an estimate, President Lyndon B. Johnson is supposed to have said, 'Ranges are for cattle. Give me a number', while another president, Harry Truman, was fed up with advisers saying 'On the one hand, this' and 'On the other hand, that' and apparently asked for a one-armed economist. Such unreasonable demands for certainty by decision-makers is not only absurd and potentially dangerous, but can also be considered as a way of shifting responsibility away from politicians on to the advisers, as in the phrase 'we are following the science' used during the Covid pandemic. I have heard horror stories of error bars being airbrushed out of graphs before they were shown to decision-makers.

In the meantime, perhaps statistician John Tukey's quote should be on the wall above every politician's desk: 'Far better an approximate answer to the right question, which is often vague, than an exact answer to the wrong question, which can always be made precise.'[11]

What is being communicated?

It may sound obvious, but we need to be clear what we are uncertain about – the *object*. For example, there is no point talking about the risk of something happening without specifying whether this is over a lifetime, next year, or tomorrow. So when the contraceptive pill is labelled at '91% effective', it may not be appreciated that this relates to a year of use, and so it's an estimate that around 9 in 100 women using the combined pill will get pregnant in a year.[12] When the US National Weather Service report a '20% probability of precipitation', they explain this is the probability at a single spot, averaged over the forecast area in the time period specified – it does not mean rain 20% of the time or over 20% of the area.[13]

Good communicators also explain the *source* of the uncertainty. Is it unavoidable unpredictability, limited evidence, doubtful assumptions in a model, or expert disagreement? This is also an opportunity to explain the incremental scientific method – at the start of the Covid-19 pandemic, saying this was a novel virus about which little was known, but careful investigation would mean that some of that uncertainty would get resolved.

The *magnitude* of the uncertainty can be expressed using words, numbers or graphics. Most conversations feature words to convey uncertainty, from saying events *might*, *may*, *could* or are *likely* to happen, to terms such as *possibly* or *perhaps*. But, as the Bay of Pigs example in Chapter 2 revealed, these words are vague and open to misinterpretation. More useful verbal alternatives include predefined categorizations, as in the IPCC's use of *likely* to mean above 66% probability; qualifying a number by saying it is an *estimate*, or is *around 30*, or that *the true value could be higher or lower*; reporting a list of possibilities as in a list of suspects in a crime; or making a claim, but acknowledging the possibility of being wrong, as in Cromwell's rule.

While words expressing uncertainty are better than nothing, the emphasis throughout this book has been on using numbers to express uncertainty and risk. For probabilities for single events, I personally avoid the word 'chance' unless it concerns games or other contexts in which probabilities can be agreed and calculated, but 'probability' can sound clumsy, so I prefer a rough frequency format such as 'about 2 out of 10 people like you', or a percentage 'about 20%'. The '1 in X' format is popular ('about one in five'), but it is best to avoid using many of these together as it requires mental effort to compare: when asked which indicates the greater risk of getting a disease – '1 in 100, 1 in 1,000, or 1 in 10?' – 28% of respondents in the United States and 25% in Germany got the answer wrong.[14] Similarly, as we saw in Chapter 12, it is best to avoid 'return periods' such as 'we expect to see this once in a hundred years', as people tend to expect a 100-year gap between such events.

A related phenomenon is *ratio bias*, when a competition offering a 9 out of 100 chance of winning is (illogically) preferred over one with a 1 out of 10 chance of winning.[15] The denominators should therefore be kept fixed so that the comparison would be between say 9/100 and 10/100, where the better choice should be more obvious. An extreme version of ratio bias occurs when the denominator is entirely ignored, so-called *denominator neglect*, such as when a single tragedy leads to demands for expensive precautionary measures, ignoring the extreme rarity of the event.

Absolute and relative risks

After coming across the clickbait headline 'Binge watching TV can actually kill you'[16] in 2013, I felt obliged to check the evidence behind this alarming claim. Japanese researchers had followed up over 75,000 people[17] for an average of around ten years each, and those watching more than 5 hours a day had 2.5 times the risk (95% confidence interval 1.2 to 5.3) of dying from pulmonary embolisms (blood clots in the lungs) compared with those watching less than two and a half hours of TV a day. The relative risk was therefore 2.5, and gave rise to the wonderful headline. But what about the **absolute risks**? It turns out that there were an extra 5.4 fatal embolisms per 100,000 years of follow-up, which means that, even if the results are true and causal, someone would have to watch more than 5 hours a night for 19,000 years to expect a fatal pulmonary embolism due to the TV watching. So maybe you don't need to cancel the streaming subscription yet.

These examples show how looking at absolute risks can bring an alternative perspective to apparently worrying stories. But there are situations where relative risks are a vital part of communication. The first concerns low-probability, potentially catastrophic events. For example, in 2009 residents of the Italian mountain town of L'Aquila were told that the absolute risk of an earthquake was low, which was true, and they subsequently did not adopt the traditional protective measure of temporarily moving out. A few days later a major earthquake killed 300 people in the town.[18] Seven Italian earthquake scientists were later convicted of manslaughter on the grounds of issuing unduly reassuring messages and not emphasizing that although the absolute risk was low, the relative risk was at least a hundred times normal. Their convictions were later quashed, but not before a chilling warning had been given about the importance of appropriate risk communication. Even though a risk of a catastrophe may be small, we all take daily low-cost precautions that make it even smaller – we wear a seatbelt and we are cautious when we cross the road.

A second context in which relative risks are important is when the absolute risks vary hugely. For example, the effectiveness of medical treatments is generally reported in relative terms, such as Pfizer reporting that their Covid-19 vaccine reduced the risk of symptomatic disease by 95% (95% interval 90% to 98%).[19] Such relative benefits are the standard way of summarizing effectiveness, and will generally be fairly constant whatever the background event rate, for example the Pfizer claim might be expected to hold in both low- and high-risk groups. In contrast, the absolute benefit will vary hugely depending on the context and the length of time being considered, being larger for say older people and when a lot of virus is circulating, and potentially being extremely small for younger people when there is not much virus around, as we shall see below.

The relative benefit determines whether a vaccine works, and so is the appropriate measure when regulatory bodies grant approval for its use. In contrast, the absolute benefits become relevant when deciding whether to take the vaccine, since they need to be traded off against any potential side effects. After all, the smallpox vaccine is very effective, but I am not going to bother to have it as there is (currently) no background risk.

––––––––

There's a particular problem in communicating and comparing small *acute* risks from sudden accidents or disasters. These are often placed on scales such as $\frac{1}{1,000,000}$, $\frac{1}{100,000}$, $\frac{1}{10,000}$, and so on, like the F-N chart in Figure 12.2. But being equally spaced on a relative-risk scale may give the incorrect impression that there are equal *absolute* risk differences – hence the changes in scale in the graphic for the UK National Risk Register shown in Figure 12.3.

Just as with any frequency format, it is better to have a fixed denominator.

Death from	Average micromorts
A skydive	8
A scuba dive (club member)	5
Commercial fishing, per day (UK)	3
Coal mining, per day (UK)	1
Travelling 7,500 miles in a train	1
Driving 7,500 miles in a car	25
Riding 7,500 miles on a motorcycle	1,000
Non-natural causes, each day (England and Wales)	0.8

Table 14.1
Average micromorts (1 in 1,000,000 risk of death) for different activities.[20]

One proposed standardization is known as the *micromort*, a unit of 1 in 1,000,000 risk of death, which allows comparisons of different activities as well as just living. Of course, these are only ballpark probabilities, and many other factors would need to be taken into account if assessing the risks for any particular individual. But Table 14.1 does help explain why I am not a motorbike enthusiast.

Our bad habits, exposures to a harmful environment and long-term diseases may not pose an acute risk of sudden death, but their cumulative *chronic* risk could shorten our life. We each have an annual hazard of dying (Figure 11.2), so in this context the relative risk is the hazard ratio. For example, eating a portion of red meat each day is associated with a hazard ratio of 1.1 (i.e. a 10% increase in annual risk of death), which would shorten life expectancy by around one year.*

We could in principle produce league tables of all the risks we face, but one of the problems with comparisons like that in Table 14.1 is that, as we saw

---

* Seeking an analogy with the micromort, I am responsible for a measure for chronic risks called the *microlife*, defined as 30 minutes' loss of life expectancy. Being 5kg overweight, or eating a portion of red meat each day, is associated with around 1 year's loss of life expectancy, which is roughly one fiftieth of your adult life. So each day with one of these risk factors is, pro rata, costing you one fiftieth of the 24 hours, which is around half an hour. I call this a microlife because 1,000,000 half-hours is 57 years, so these exposures are, on average, losing you one millionth of your adult life. Sadly this measure hasn't become popular.

in Chapter 1, people may have very different emotional reactions to potential threats – a voluntary risk from an enjoyable activity feels very different from one imposed on us by our job. We should ideally keep the context, and so the feeling associated with the risk, constant, and so standard yardsticks such as the chance of being struck by lightning, or throwing so many sixes in a row, are generally unhelpful. Instead, when, say, communicating the risk from earthquakes, comparison should be made with the risks at other places on earth susceptible to earthquakes.

There are some other brief points to remember if you ever want to communicate numbers summarizing uncertainty and risk. First, '98% survival' sounds better than '2% mortality', demonstrating that the *framing* of the numbers can influence people's feelings, and so it is best to provide percentages or frequencies of both positive and negative outcomes. Second, uncertainty about quantities can be expressed as a full probability distribution, as in Bayesian posterior distributions (Chapter 7), an estimate and a 95% interval, or a rounded number or a range, as in the National Risk Register. Again, the framing of intervals can be important. The 2009 UK climate projections assessed an uncertainty interval for a possible increase in temperature which had a top end of 12°, and one aim was to avoid the media saying the change 'could be as high as 12°'. So the projections gave the true value as 'very unlikely to be greater than 12°', and this clever shift from a negative to a positive frame changed the tone of the coverage.

Figure 14.1 shows some of the many ways that uncertainty about quantities can be visualized. Error bars are extensively used, for example to show the uncertainty intervals for the estimates of R (Chapter 8), yet a problem is that they draw a sharp distinction between values inside and outside the interval, and can encourage the incorrect impression that values in the interval are equally likely, and those outside are essentially impossible. My preference is for violin or gradient plots, which avoid hard boundaries (the gradient plot has density of ink proportional to the probability). Icon arrays are popular ways to show, for example, what we would expect to happen to a hundred similar people, although we have found that human icons can be too emotive and may be better replaced by blobs, and that colours representing poor outcomes and death should not be too glaring.*

---

* In principle, uncertainty about the number of icons could be represented by some shading, in which case it may be better to have the 100 icons in a single line, like a horizontal bar.

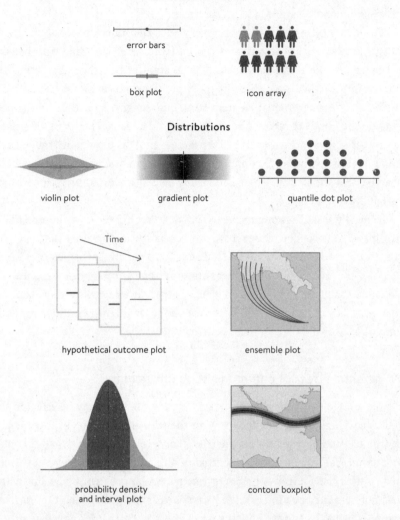

Figure 14.1
Some techniques for uncertainty visualization. Adapted from Padilla, Kay and Hullman.[21]

Figure 14.1 includes what Padilla and colleagues call a 'contour boxplot', representing, say, the possible path of a hurricane across some territory. This uses a limited set of shaded uncertainty intervals and so is analogous to the Bank of England's fan chart. This is perhaps my favourite visualization, preferably without the central estimate, as it manages an impression of uncertainty and remains simple and interpretable.

There has been a lot of interest in visualizing uncertainty about hurricane tracks, since this can be a crucial issue when deciding whether to evacuate or take other precautions. The standard method is a 'cone of uncertainty', which is a 66% probability prediction interval for the path of the hurricane, but which, like error bars, may suggest to people that areas outside the cone are 'safe'. Alternatives include a 'spaghetti plot' of simulated paths, or a more structured 'ensemble plot' as shown in Figure 14.1. An animated 'hypothetical outcomes plot' shows possible realizations of a Monte Carlo simulation, which can vividly display uncertainty and potential variability in outcomes.

Ironically, visualizations can be too good. If they are too engaging and realistic, people may believe they actually represent the truth, rather than being a construction based on models and judgement. As we saw earlier, Bank of England fan charts may not properly communicate the 10% of deeper, unmodelled uncertainty in the tails. A challenge is therefore to make a visualization that is attractive and informative and yet conveys its own contingency and limitations.

What is the effect of communicating uncertainty?

This is a huge and very complex issue, since audiences may be affected in what they think, what they feel or what they do, and in any case there is an enormous variability in how people react. There is no doubt that many people find numerical expressions of uncertainty difficult to understand, but this is tied up with general lack of numeracy and particular problems in dealing with low numbers such as 1 in 1,000,000. Of course, standardization of terms like 'likely' is valuable, and we know that positive or negative framing can affect understanding of magnitude. People also vary in how they respond emotionally to uncertainty; for example when shown a range around their risk of developing cancer, those who were naturally more optimistic expressed less worry, perhaps as they could focus on the bottom end of the range.[22]

There is limited research about the effects of communicating 'indirect uncertainty' based on the quality of evidence (Chapter 9). Experiments suggest that people downweight their confidence appropriately when told that a claim is based on low-quality evidence, although if people are not told about the quality of evidence, they tend to assume it is good. This demonstrates a touching form of trust,[23] which unfortunately can be abused by people who make spurious claims based on low-quality evidence.

Evidence suggests that confident expression of uncertainty, say by using ranges, does not in general reduce trust in the source of the message.[24] This is reassuring – it would be very unfortunate if being honest enough to acknowledge uncertainty meant that people spurned our opinion and turned to some charlatan who expressed complete confidence.

What should we expect from trustworthy communication?

At the start of this chapter I emphasized the need to demonstrate trustworthiness, but how can we best do that? I was part of a group that came up with five brief principles, in which uncertainty plays a major part.[25] If we want to be trustworthy, we should seek to

1. *Inform rather than persuade*: In general, the basic ethos of trustworthy communication is to empower audiences to make decisions that fit with their values, rather than manipulate them into doing or thinking what the communicator wants. Although there may be crisis situations when persuasion is appropriate.
2. *Be balanced*: Feature both pros and cons, benefits and harms, winners and losers. Although not false balance; disputes about climate change are not 50:50.
3. *Be upfront about uncertainty*: By using all the ways described in this chapter, whether verbal, numerical or graphical.
4. *Acknowledge limitations in the evidence*: By using the ideas about indirect uncertainty in Chapter 9, we can be clear about the quality and strength of the evidence.
5. *Try to pre-empt misunderstandings*: This requires knowing your audience and how your evidence may be either misinterpreted or misused by others. For example, UK statistics regulator Ed Humpherson has recommended that producers of official statistics emphasize what can and cannot be concluded from the data.[26]

These principles have now become part of the UK government's RESIST-2 counter-disinformation toolkit,[27] but it is reasonable to ask what the effect of this approach might be – it would be unfortunate, to say the least, if all this trustworthiness was met with scepticism and suspicion. So my colleagues carried out large studies in which messages with different formats ('persuasive' or 'balanced') about Covid-19 vaccines or nuclear power were shown to over 1,000 people,[28] each of whom was randomly allocated to see one message.

For people who were already accepting about Covid-19 vaccines or nuclear power, there was no difference in the extent that they thought the messages were trustworthy. But for people who were sceptical, the 'balanced' message was judged considerably more trustworthy than the 'persuasive' format. This is an important and encouraging finding, and means that the standard way that many parts of government communicate – trying to persuade audiences to believe something – is actively *decreasing* trust in the very group they are most anxious to reach, compared to what could have been achieved with a non-persuasive format. Which could be an important lesson for everyone communicating.

———

With my colleagues Alex Freeman and John Aston, I had an opportunity to put these ideas into practice when we were asked to help communicate the benefits and harms of the AstraZeneca Covid-19 vaccine in April 2021, after reports of serious blood clots and amid growing concern about the vaccine. We were sent data on the clots, and fitted a smooth line for the risk at each age, since age is such an important determinant of both the benefits and harms of vaccination. We then compared these potential harms with benefits of similar importance, in this case avoiding admission to intensive care. As emphasized above, the absolute benefits depend crucially on how much virus is circulating, and we prepared a range of scenarios – Figure 14.2 shows the 'low exposure risk' present at that particular period in 2021. As discussed above, the analysis uses absolute rather than relative benefits, since these are the relevant measures when deciding whether to offer, or accept, the vaccine.

In relation to our five principles outlined above, we were mainly focusing on the first two – informing rather than persuading, and being balanced. So the communication was not trying to say vaccines were 'safe and effective', but rather demonstrating that they may be safe and effective *enough* to give to some people in some circumstances. We did not introduce uncertainty into the plot, as this would have meant adding undue complexity by, say, blurring the dots, and would not have added to the message. The quality of

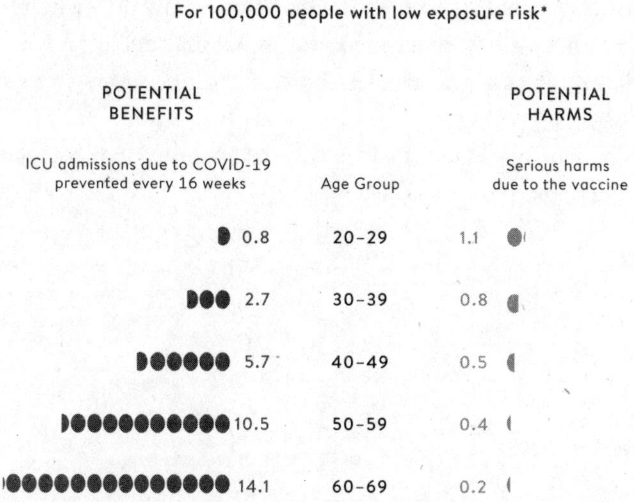

For 100,000 people with low exposure risk*

POTENTIAL BENEFITS		POTENTIAL HARMS
ICU admissions due to COVID-19 prevented every 16 weeks	Age Group	Serious harms due to the vaccine
● 0.8	20–29	1.1 ●
●●● 2.7	30–39	0.8 ●
●●●●●● 5.7	40–49	0.5 (
●●●●●●●●●● 10.5	50–59	0.4 (
●●●●●●●●●●●●●● 14.1	60–69	0.2 (

Figure 14.2
Estimates of the major benefits and harms of the AstraZeneca
Covid-19 vaccine, as used in a BBC broadcast on 7 April 2021. For
older people the benefits clearly outweigh the potential harms,
but for younger groups these become finely balanced.

*Based on coronavirus incidence of 2 per 10,000:
roughly UK in March

our evidence was reasonably good, and people tend to assume this anyway,
although we later added in some caveats on the graphic that there were
other potential benefits and harms that we were not quantifying. Finally,
we did not highlight potential sources of misunderstanding on the graphic
itself, and have been pleased to see that it has not been misused.

Jonathan Van-Tam, the respected and trusted Deputy Chief Medical
Officer, took considerable time to explain this complex infographic to the
public, and then concluded by saying the vaccine was no longer going to be
recommended for people under thirty (this was later raised to forty). This
was accepted, there were no accusations of a U-turn, no strong concerns
about the vaccine, and the representation was copied extensively.

Although the graphic could be considered as communicating the rough
chances of benefits or harms to an *individual* in the future, it was primarily
used to explain the reasoning behind a policy to be applied to a *population*.

Which naturally brings us to the vital question – how do we handle decisions under uncertainty, whether as an individual deciding whether to get vaccinated, or as a government body deciding who is to be offered the vaccine?

This is challenging, which is partly why decision-making has been left to the end of the book.

Summary

- It is vital to communicate uncertainty clearly, but this is not straightforward.
- Rather than aiming to be trusted, try to demonstrate trustworthiness.
- In a crisis, give guidance but acknowledge uncertainty and provisionality of advice.
- Identify audiences' needs, beliefs and skills, and evaluate communications.
- Choose metrics and visualizations carefully so as not to mislead people.
- Use multiple levels and formats of communication, for a range of numeracy skills.
- Try to inform, be balanced, acknowledge uncertainty and limitations of evidence, and pre-empt misunderstandings.
- Evidence suggests that being trustworthy leads to an increase in trust in those who are more sceptical.
- There is no point in being trustworthy if you are dull – so the aim is to be vivid and engaging, but without being manipulative.
- It is not enough for communicators to summarise their analytic uncertainty – they should try and ensure that the audiences end up with an appropriate impression of the reliability of any claims.

CHAPTER 15

Making Decisions and Managing Risks

Life is one long series of decisions made under uncertainty. Most of these we make without much thought – what time to set off for an appointment, or what clothes to wear when we go out in iffy weather. Some more important decisions may lead us to pause and think a bit more slowly – where to go on holiday, or which car to buy. Others may be really crucial – whether to have children, whether to run away and join the circus, or what treatment to choose for our cancer.

In theory, using ideas first developed by Frank Ramsey (Chapter 3), there is a formal mechanism to decide the best thing to do, with the following four basic steps:

1. Construct a list of possible actions, and the possible consequences of those actions.
2. Assign a probability for each possible consequence, given each action.
3. Give a value to each of those possible futures.
4. Take the action which maximizes the expected benefit.

These steps form the basis for the economic concept of 'rational' human behaviour, embodied by the ideal *Homo economicus*.

This may seem a reasonable structure, and we might follow these rules in what decision-theorist Leonard Savage called a *small world*[1] – a controlled situation such as gambling at roulette. But the real, large world is a much messier place, and it is a tall order to specify all the possible actions, out-comes, probabilities and values, as we discussed in our chapter on deep uncertainty. And even if we could carry out steps 1 to 3, making a decision might be somewhat more complicated than maximizing expected benefit.

Let's consider the most basic example. Figure 15.1 shows what is known as a *standard gamble*, which is essentially a trade-off between a guaranteed out-come and a good or bad result chosen at random.

The expected reward if we gamble is p Value(win) + $(1 - p)$ Value(lose), and so according to the rules listed above, we should reject the gamble and accept the sure thing if

$$\text{Value(sure thing)} > p \text{ Value(win)} + (1 - p) \text{ Value(lose)}.$$

Suppose you are to flip a fair coin, with heads winning £1, tails you get noth-ing. Then the expected return is ½ × £1 + ½ × £0 = 50p, so if offered 50p as a sure thing, you should be indifferent about whether to gamble or take the 50p for certain. But would you be indifferent, or would you prefer the 50p for certain?

And what if the stakes were raised? Now imagine you're on a game show and have already won £5,000 – call this Game 1. You now have an option; (a) accept an extra £10,000 for certain, or (b) try to guess a coin flip, winning £20,000 if you are right, nothing if you are wrong. What would you do – sure

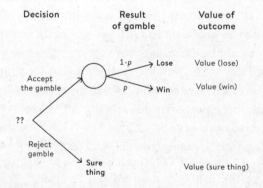

Figure 15.1
A 'standard gamble' – the choice is between choosing a 'sure thing' and receiving Value(sure thing), or accepting the gamble. This has probability p of winning and receiving a reward worth Value(win), and probability $1 - p$ of losing and getting something worth Value(lose).

	Small probabilities	Large probabilities
Gains	**Risk-seeking** over low-probability gains, such as preferring to buy lottery tickets for the 'opportunity'	**Risk-averse** over high-probability gains, such as preferring a sure gain rather than 50:50 on doubling gain – 'a bird in the hand is worth more than the chance of two in the bush'
Losses	**Risk-averse** over low-probability losses, such as preferring to buy insurance to hedge against disasters	**Risk-seeking** over high-probability losses, such as preferring 50:50 on doubling loss to sure large loss

Table 15.1
Common behaviour when making decisions under uncertainty which do not fit the standard model for rational decision-making.

thing or gamble? I think most people would choose the sure thing, unless they had got a bit carried away.

If you chose the sure £10,000, you are demonstrating 'risk aversion', in that you avoided the gamble, even though the expected gain is £10,000 under both options. But what if the prize in the gamble were £40,000, while the sure thing was stuck at £10,000? Would you feel differently? Risk aversion only goes so far; at some point, you would presumably take a chance.

One way of thinking about these sorts of situations, first developed by Daniel Bernoulli (yet another nephew of Jacob Bernoulli) in 1738, is to set the value – or *utility* – of money to be non-linear. In other words, the value we place upon each unit of money decreases as the amount goes up – getting the first £1,000 is worth far more to us than a change from £19,000 to £20,000. This means that the utility we place upon £10,000 for certain is greater than half the utility of £20,000, and shows why people tend to be risk-averse for higher-probability gains.

Now let's change the story of the game show – call it Game 2. Try to imagine that you have already won £25,000, and the cash is in your pocket and you're getting ready to go home, feeling rather pleased with how things have turned out. But then they call you back for a final twist: you need to choose between (a) giving back £10,000 for certain, and (b) guessing a coin flip, and if you lose you have to give back £20,000, and if you guess correctly you get to keep all your winnings. What might you do? It turns out that many people say they would choose to gamble rather than face a certain loss. This is known as being risk-prone or risk-seeking for higher-probability losses.

But a little analysis shows that both Game 1 and Game 2 are identical in terms of their final outcomes; (a) £15,000 for sure versus (b) 50:50 chance of £5,000 or £25,000. This shows that our choices under uncertainty may not follow an apparently 'rational' structure – it's not only where you end up that matters, but also where you start from.

It all gets even more complicated when we consider small probabilities. People buy lottery tickets, even though the expected gain is much less than the price of the ticket, so they are being risk-seeking for low-probability large gains. But we buy insurance, paying more than the expected loss (otherwise insurance companies would go bust), so we are risk-averse when there is a small probability of a large loss.

This gives the classic fourfold pattern in Table 15.1 identified by psychologists Daniel Kahneman and Amos Tversky,[2] showing common behaviour when faced with decisions under uncertainty with known outcomes and probabilities.

In response to these observations, Kahneman and Tversky came up with *prospect theory*, a more complex mathematical framework for decision-making under uncertainty, explaining aversion to losses, overweighting low probabilities and focusing on changes rather than final states. The basic ideas of prospect theory have been empirically confirmed,[3] although Kahneman has said it is still not intended to fully describe human behaviour,[4] which is much more subtle than can be captured in a formula.

––––––––

So far we have unrealistically assumed that we know all the probabilities and outcomes, but consider the following choice that I give to school students. I have two bags:

- Bag A has 5 red and 5 black balls.
- Bag B has 10 balls, all either red or black, where the number of red balls has been picked at random from 0 to 10.

You need to choose a colour, red or black, then a bag, A or B, and then a ball from your bag. If you pick a ball of your chosen colour, you win a prize. Would you rather pick Bag A, with a known chance of success, or Bag B, with an unknown chance?

In terms of expected gain, the options are identical; there's a complete symmetry between the two colours, and so it cannot pay to choose any combination of colour and bag rather than another. But people tend to prefer Bag A, the one with the known 50:50 chance. This risk aversion to uncertainty about probabilities is called *ambiguity aversion*, based on pioneering work by economist and activist Daniel Ellsberg* in 1951.[5]

If you think back to the discussion about 'confidence' (Chapter 9), we saw that analysts were unwilling to make confident judgements when they knew they were missing a vital piece of information that could have a dramatic influence on their opinion. If we choose bag B, with the unknown share of red and black balls, we are exactly in this 'information gap', so people prefer to avoid this and go for a high-confidence situation where we know the odds. We could consider this a robust strategy, where we won't be kicking ourselves later. Or as a pessimistic strategy, in which we assume that whatever we choose, the worst will happen, and take the action that guarantees the least-bad result.

Of course, in most real-life decisions we have more than just ambiguity about the probabilities; we also don't know the consequences and how we will feel about them, and may not even be aware of all the options available – we have deep uncertainty, as we explored in Chapter 13. No formal theory can deal with low-probability high-impact events where both the risks and the outcomes are poorly understood. And crucially, we rarely are faced with a single one-off irrevocable decision, since usually there is a whole sequence of minor judgements, possibly leading to commitment to a path without having ever sat down and positively decided this was the best thing to do. You may be able to recognize this occurring in your life.

––––––––––

So, if the formal theory is challenging, how are we supposed to make decisions in the face of all this uncertainty? There seem to be four broad strategies, representing a broad continuum of decreasing technicality, which can be applied to both decisions made by individuals and those on behalf of organizations or governments.

––

* In 1971 Daniel Ellsberg released the *Pentagon Papers* to journalists, revealing the US government's dishonesty about the Vietnam War. The Nixon administration then commissioned the 'White House Plumbers' – later to commit the Watergate break-ins – to burgle Ellsberg's psychiatrist to find embarrassing information, and illegally tapped Ellsberg's phone. This was all revealed at Ellsberg's 1973 trial, at which he was acquitted of charges of espionage, theft and conspiracy.

1. *Full decision analysis*, as laid out above. This will only be realistic in a fairly controlled small-world situation in which it can be assumed that options, probabilities and values can at least approximately be fully quantified, for example a string of gambles. The previous examples have shown that we should not always expect the conclusions to match human intuition. As another (admittedly wholly unrealistic) example, suppose you had a fixed amount of money which could be spent on treating just one out of three groups with diseases of different severities;

    (a) Group with disease A: Out of every 100 people, all would normally die, but you could save 3 of their lives.

    (b) Group with disease B: 50 of every 100 people would normally die, and you could save 3 of their lives.

    (c) Group with disease C: 3 out of every 100 people would normally die, and you could save all of their lives.

    Each of the options leads to the same total benefit in terms of lives saved. But you might feel that option (a) at least gave some hope to people who would definitely die, while option (c) cured an occasionally lethal disease. Option (b), which just reduces risk by a small margin, may not seem so attractive.

2. *Semi-quantified analysis*, in which we do our best to list options, and even judge probabilities and values, but fully acknowledge the limitations and look for strategies that are robust to all the things we know we don't know. When the team at the Winton Centre for Risk and Evidence Communication works on NHS decision-aids for patients, we present some rough estimates of rates of people recovering and side effects of alternative options.[6] There is no formula guiding someone to a decision, but the aim is to fully consider all options, explore feelings, and encourage patients to be as precautionary or bold as they wish. Success would mean that afterwards, whatever happens, the patient at least felt they had full information when the decision was made.

3. *Heuristic*, where decisions are reached by using informal and unconscious rules of thumb. Psychologist Gerd Gigerenzer has popularized the idea that many decisions in the face of uncertainty are made rapidly and ignore much of the information available – so-called *fast and frugal* strategies;[7] for example, if asked which of a pair of cities is the bigger, just choose the most famous one. These can work well in daily life, but of course examples can be contrived in which they fail.

4. *Story-based*, or imagining possible futures. This may be close to what we routinely do – Daniel Kahneman is supposed to have said, 'no one has ever made a decision because of a number, they need a story.'[8] And if we allow ourselves to ruminate on gloomy narratives, then we may be led naturally to precautionary behaviour, trying to make us resilient to the worst that could happen. These ideas have been expanded into Conviction Narrative Theory[9] for making decisions under deep uncertainty, in which people focus on a narrative that feels 'right' to explain the available evidence, use that narrative to imagine possible futures, and non-numerically judge the value of those imagined futures to make a choice. While this may be fairly descriptive of what we do in our daily life, personally I am unconvinced that emotionally driven convictions are an appropriate basis for making serious decisions – we get enough of this on social media. I believe it is better to encourage people to think slowly and assess magnitudes where they can – of course without ever believing any of their assessments are 'correct'.

There have been whole books written comparing the first with the third and fourth strategies, often criticizing risk assessors, economists and financial analysts who actually seem to believe their models. But this seems a false split between two extremes, rooted in Frank Knight's outdated distinction between quantifiable 'risks' and unquantifiable 'uncertainties' (Chapter 13).

In contrast, the rest of the chapter mainly illustrates the second strategy, to quantify as much as is reasonable, while trying to be aware of the inevitable inadequacy of any analysis. And we start with some of the trickiest things to put into numbers – government policies that involve risk to humans, and that are both expensive and contested. It is natural to add up expected costs of policies that affect whole societies, but quantifying the benefits of government interventions is more challenging, particularly when they may save (or lose) lives.

Policy decisions

The UK Treasury's Green Book[10] gives guidance on appraising policy options using either *cost–benefit analysis*, in which benefits are given monetary value, or *cost-effectiveness analysis*, in which the cost to achieve a unit of benefit is compared; in both cases future costs and benefits will be discounted by a fixed amount per year. For example, in considering possible road improvements, a monetary value can be placed not only on time saved but on any projected reduction in road casualties. This requires a value of preventing a fatality (VPF), which the UK Department for Transport revises every year and is currently over £2 million.

This also enables apparently cold-blooded assessments of whether additional safety measures are worthwhile. For example, in the 1990s an Automatic Train Protection system was being considered that would constantly monitor, and if necessarily control, train speeds, but it was cancelled when it was estimated that it would cost £9–10 million for each fatality prevented, while the VPF was then only £700,000.[11] Such considerations can be overridden when there is high societal concern; vast amounts are spent on nuclear-waste disposal, regardless of any actual risk to human health.

Transport policy seems straightforward compared to the natural environment. How would you put a value on what forests, woodlands and trees provide to society? The Office of National Statistics has to do this for 'Natural Capital Accounts', estimating that the total annual value of woodland in the UK in 2020 was an estimated £8.9 billion. About half of this was due to carbon capture, but over £1 billion is expenditure from 800 million tourism and recreation visits and a further £1 billion from health benefits.[12] But cultural and spiritual value, say from ancient trees, is not currently monetized.*

Of course, all these assessments of costs and benefits are rife with uncertainty, and the Green Book requires 90% intervals for outputs in cost–benefit or cost-effectiveness metrics, and recommends the sort of extra detail we have emphasized throughout this book, such as sensitivity analysis to assumptions, identifying important factors that drive the conclusions, and non-monetizable benefits that are not included. 'Business as usual' must always be an option, and a precautionary allowance has to be made for 'optimism bias'; the recommended adjustment for non-standard civil engineering projects is to allow up to 66% cost overruns, reflecting bitter experience of costs going way beyond the calculated uncertainty intervals. There are claims that, for mega-projects, naive optimism tends to be dominated by 'strategic misrepresentation' – the deliberate understating of costs for political reasons.[13]

---

* The cultural importance of trees was emphasized when a famous solitary sycamore on Hadrian's Wall was deliberately cut down in September 2023, provoking widespread distress and expressions of loss.

The value of human existence is not just a matter of length of life, but also the quality of life. The UK National Institute for Health and Care Excellence (NICE) has for decades carried out cost-effectiveness analyses to help decide which treatments will be paid for by the National Health Service (NHS), based on estimates of the cost to achieve an additional *quality-adjusted life year*. This requires assigning values to medical conditions so, for example, in the EuroQol 5D scale used in many appraisals, 'severe anxiety or depression' is judged to subtract 0.29 from your annual quality of life, meaning a year with this condition is only 'worth' 71% of a healthy year.[14]

These values are generally obtained from surveys of the general public using *time trade-offs*, so that presumably respondents judged that, on average, 5 years in full health is worth 7 years with severe anxiety or depression, leading to a utility value for this condition of $5/7 = 0.71$. Although these values may be reasonable when evaluating impacts on groups of people, it is not clear that they would work as individual 'utilities' in the standard decision-theoretic framework. If we assign healthy life as 1, and death as 0, a utility of 0.71 would mean that someone would, in theory, be willing to accept an operation that would cure their depression if successful, but had a 29% mortality rate. I have no idea how acceptable this would be.

––––––––––

In the last chapter we saw how our infographic (Figure 14.2) was used to explain the policy decision about which age groups should not be recommended to get the AstraZeneca Covid-19 vaccine and, although its scope was limited, we felt it clearly illustrated the benefit–harm trade-off in the future.

But another feature becomes apparent when we look back on this analysis. There is a qualitative difference between those being harmed (the paler dots on the right) and those benefiting (the darker dots on the left), in that those who are harmed become identifiable people, potentially given names and faces, while those benefiting are 'statistical' people – nobody ever knows who has benefited from a vaccine.

As anyone on social media will be aware, there can be strong societal reactions against the imposition of harm from vaccinating healthy people, no matter how rare the side effects. Indeed, people harmed by this and other vaccines against SARS-CoV-2 have (at the time of writing) begun legal cases against the manufacturers. This issue also arises with 'smart' motorways in which the hard shoulder is used as an extra lane; these may save statistical lives, since the improved traffic flow encourages more people to use motorways rather than more dangerous A-roads, but this can come at a cost of some highly identifiable victims of accidents.

This illustrates that, when it comes to policy decisions, it is vital to understand and take into account *societal concern*. There is a major difference between the perception of statistical and identifiable lives – it is likely that far more than the Value of Preventing a Fatality of £2 million would be spent to save the life of a specific individual who has attracted media attention, say a child stuck down a well. And, of course, societal concern becomes particularly relevant when developing regulation to try to reduce the risk of harm to the public.

Regulation and risk

We want to be protected from hazards, but we don't want our freedoms curtailed. We want 'safe' products and a clean and sustainable environment, without damaging the economy. How can we balance these competing demands?

A substantial 'risk' industry has grown up around developing regulations and guidelines for what is allowed in organizations or in our society. This falls broadly under the umbrella of *risk governance*, involving a process of risk analysis, communication and management, ideally with public and stakeholder engagement at each stage. Some institutions try to separate these roles, which makes it even more important that uncertainties are properly communicated using all the verbal, numerical and graphical tools that we've seen throughout this book.

Risk-analysis strategies will generally be semi-quantitative, with the role of formal analysis getting less as the uncertainty gets deeper. But when it comes to serious risks at work, at least one major organization has been prepared to put their judgements into numbers.

> **What is an acceptable risk of being killed at work?**

In 2001 the UK Health and Safety Executive (HSE) released a highly influential document called 'Reducing risks, protecting people'.[15] *R2P2*, as it became known, took an innovative approach to health and safety at work; crucially, it makes no mention of making anything 'safe', and instead puts all threats in relation to *acceptable* and *intolerable* risks. This became known as the *Tolerability of Risk* framework.

Figure 15.2 shows the HSE's approach to risks to individuals from workplace accidents. A 1 in 1,000,000 chance of an employee being killed at work each year is considered *broadly acceptable* – this does not mean safe, but safe enough. But a 1 in 1,000 chance is considered *intolerable* for a worker, as is 1 in 10,000 for a member of the public. Coal miners and commercial fishermen (see micromorts in Table 14.1) are two of the occupations that often feature in the intolerable zone. If the estimated risk lies between the intolerable and broadly acceptable regions, then this is only tolerable if the risks are made *As Low As Reasonably Practicable* (ALARP), meaning that risk-reduction measures should be adopted if they are proportionate.

Looking beyond the experience of individuals, major industrial accidents can lead to mass fatalities – I still remember my shock when the chemical works in Flixborough exploded in 1974[16], but the twenty-eight deaths on that day were dwarfed by the 1984 escape of a poison-gas cloud from the Union Carbide works in Bhopal, India, which killed over 2,000 people and injured tens of thousands more.[17] The impact of these events is more than just adding up outcomes for individuals; societal concern reveals itself in public outrage, and political and media reaction. This is not readily measurable, but the HSE

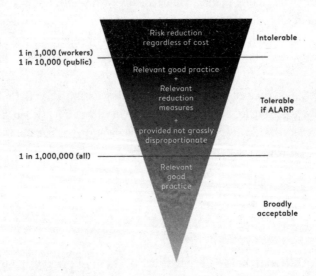

Figure 15.2
The tolerability of risk framework for individuals and workplace accidents, as pioneered by UK Health and Safety Executive.

rather boldly said that an industrial installation with a 1 in 5,000 annual chance of causing 50 deaths is intolerable, but a 1 in 500,000 chance (a hundredth of the risk) was broadly acceptable. These limits could be extended proportionately, so that a 1 in 500 annual chance of causing 5 deaths is also intolerable.*

HSE says that both the individual risks of the most exposed people, and then the societal risks, need to be assessed, and then 'only when both types of risk are demonstrated to be tolerable and ALARP can the duty of the operator be considered met.' This puts a heavy responsibility on those building models for possible failures, as we have seen how such models for low-probability high-impact events are particularly prone to over-precision, limited scope, uncertainty and errors. No analysis can claim to be 'correct', and so the robustness of the conclusions to alternative assumptions is essential. Particularly if you are operating a plant that could explode.

Toxicology and environmental exposures

While few of us live close to high-hazard industrial installations, we all consume food and breathe air, and we would prefer not to be harmed in the process. Regulators have the job of setting maximum recommended exposures for everyday hazards such as pesticides and food additives, as well as chemicals used in industrial processes. Uncertainty is fundamental to this quantified, but precautionary, process.

The basic idea is to carry out animal experiments, generally involving mice which have been specially bred to be susceptible to developing tumours, to determine the highest exposure that either has no observed adverse effects, or does not lead to undue extra risk. This tolerable level for mice is then divided by *Uncertainty Factors* (UFs) to set exposure limits for ordinary humans. The idea of Uncertainty Factors, also known as 'margins of safety', goes back to the 1950s, when they were originally set at 100, i.e. the tolerable dose in animals was divided by 100 to apply to humans. The current default way of dealing with the extrapolation from mice to everyday human life is to use separate uncertainty factors for transferring the exposure from animal to human, from average human to sensitive human, and from short-term to long-term exposures – these are then multiplied together to give an

---

* Technically, this means that regions of broadly acceptable and intolerable societal risks are delineated by lines with gradient -1 on an F-N diagram (Figure 12.2).

overall margin of safety. These are like engineering safety factors when building bridges, but rather larger.

Of course, none of us want to be harmed by the food we eat, but sometimes precaution might be overdone.

How much burnt toast is it OK to consume?

In January 2017 the UK Food Standards Agency launched its Go for Gold campaign, encouraging people to avoid burning their food. Its reasoning was that acrylamide, a chemical that forms on burnt food, has been labelled by IARC a 'probable carcinogen' (see Chapter 10). The FSA provided no estimate of the current harm caused by acrylamide, nor the benefit from any reduction due to people following its advice, but the campaign led to headlines such as 'Browned toast and crispy roast potatoes "a potential cancer risk"'.[18]

Our team* received an embargoed warning of this campaign, and as I really like crispy roast potatoes, I felt we should critique the evidence of potential harm. First we noted that intensive efforts have failed to show good evidence of any quantifiable link to cancer in humans and, as we saw in Chapter 10, IARC was describing a hazard rather than a risk.[19] Second, mice experiments had estimated the benchmark dose of acrylamide to be 170µg/kg-body-weight/day – this means we can be confident that exposures below this level will not cause a measurable increased risk of tumours in mice.†

Compared to this benchmark level of 170 in mice, even high human consumers of acrylamide, say eating a slice of burnt toast a day, get through only 1.1µg/kg/day. This is only a 160th of the benchmark level in mice, which might seem to be rather reassuring, and may explain why it's been so difficult to observe any effect of acrylamide in human diet. But because this hazard concerns cancer, toxicology committees demand a rather arbitrary margin of safety comprising two Uncertainty Factors of 100, in other words the acceptable dose in humans is set as one 10,000th of that for mice. That means that an acceptable exposure is about one 60th of the consumption of high-consuming adults, meaning people should only eat a thumbnail-size piece of burnt toast

---

* At the Winton Centre for Risk and Evidence Communication in Cambridge.

† Formally, the benchmark is the lower end of a confidence interval for the dose that would cause tumours in 10% of susceptible mice.

every day, in spite of the lack of actual evidence for human risk. This was the basis for the FSA's campaign, and we were not impressed.

Ridicule in the media, some of it due to our intervention, led to a rapid cancellation of the Go for Gold campaign, but I have since heard of individuals who are obsessively anxious about burnt food. Perhaps more worrying, people may just consider this yet another scare story from scientists and lead them to dismiss truly important warnings about, say, the potential harms from obesity.

Acrylamide is also produced when roasting coffee, and in 2018 a California court ruled that signs should be posted at coffee-shop entrances warning about a possible link with cancer. At the time, I used US Environmental Protection Agency data to estimate that a daily cup of coffee could increase your absolute lifetime risk of cancer by 0.0003%, and as around half of us will be diagnosed with cancer at some point, this does not seem a very important addition to 50%.[20] The California Office of Environmental Health Hazard Assessment (OEHHA) came to similar conclusions, and in 2019 decided that no warnings need to be given as the acrylamide dose in coffee was too low to be a carcinogenic risk.[21] In all this argument it tended to get overlooked that there is good evidence that coffee is beneficial to health, and is indeed linked to a *reduction* in cancer risk.[22] So one lesson is that we need to take a broader view of both the potential harms and benefits of any activity, rather than obsessively focusing on specific harms.

The default use of Uncertainty Factors to set acceptable limits to exposures seems rather crude, and although there are claims the procedure is not as conservative as first appearances might suggest,[23] there is continuing effort to make these less arbitrary, and construct proper models for the actual risks to humans.[24] For me, the real problem is having a single threshold for an 'acceptable' exposure, which encourages the perception that anything above this is 'unsafe', which is nonsense. This shows the huge value of HSE's Tolerability of Risk framework in having *two* thresholds, a high one delineating intolerable risk that must be avoided, and a low one representing a broadly acceptable chance of harm. This was a bold but sensible innovation in the delicate area of 'health and safety'.

Being precautionary

There are numerous examples where official bodies were slow to recognize potential threats. For example, there were warnings about the potential harm from X-rays and other forms of ionizing radiation from the early 1900s, but it took decades for adequate protection to be taken; when I was young, X-rays were still being used routinely in shoe shops to check children's fittings. Similarly, deaths of asbestos workers became noted before 1910, with evidence of harm growing over the decades. There was some inadequate regulation, but it took nearly a hundred years for a ban; in 1999 it was estimated that around 250,000 people in western Europe would die of mesothelioma over the following thirty-five years.[25]

Such past failures have led to recommendations of a general *precautionary principle* regarding risks to society. This has many different versions; a particularly strong one states that no action should be enacted unless it is provably 'safe', which seems more paranoid than precautionary, although this was the approach in the 2018 Californian court case against coffee retailers – they had to prove their product was safe, and so were guilty until proven innocent.

A weaker precautionary principle, as widely promoted by the European Union, states that we should not wait for conclusive scientific evidence before protecting against a possible risk. This may at first seem very reasonable in the face of growing threats to the environment – we should be just like the final little pig and build our house of bricks, even though we don't know for certain if a big bad wolf will come along and huff and puff.

But why stop at bricks? Why didn't that little pig build a shelter capable of surviving a nuclear winter? There has been a reaction against excessive enthusiasm for precaution based on worst-case scenarios – Philip Tetlock has said, 'We cannot go full throttle on pre-empting every candidate threat that pops up on someone's radar screen. We must set priorities.'[26]

Precaution can have unintended consequences. Germany has a long-held aversion to nuclear power, reinforced after the Chernobyl disaster, and it has been finally phased out after Fukushima. This means they have increasing reliance on coal-powered plants, which further contribute to climate change. And this is not the only way that the Fukushima event caused indirect harm through an excess of precaution. We have already seen (Chapter 13) how this nuclear power plant in Japan was built in a known tsunami zone and yet was not adequately protected against extreme waves. Then the 2011 tsunami struck, the cooling systems failed, and radioactive contaminants were released. There was huge international interest, with the European Union's energy commissioner, Günther Oettinger, announcing, 'there is talk of an apocalypse and I think the word is particularly well chosen.' Over 150,000 local residents were evacuated, and the incident was put on level 7 on the International Nuclear and Radiological Event Scale (INES) – this is the highest point on the scale, achieved previously only by Chernobyl.[27]* But was this an appropriate response to the risks?

While the tsunami killed at least 18,000 people, no Fukushima workers died (although one fatal lung cancer was later attributed to the radiation). But the evacuation caused massive social, economic and mental health harms, including over fifty immediate fatalities of inpatients and elderly people during the evacuation, while nearly 1,800 subsequent deaths were classified as 'disaster-related'.[28] Excessive precaution, spurred on by fears of radiation, led to measures that produced far more harm than benefit.

———

So what can we do when making decisions in the face of uncertainty? My personal conclusion is that it is best to work down the list of strategies outlined earlier, trying to quantify uncertainty, but acknowledging when this is inadequate. And as we increasingly acknowledge deeper, ontological uncertainty, where we don't even feel confident in listing what could happen, we move away from attempts at formal analysis and towards a strategy that should perform reasonably well both under situations that we have imagined, and those we haven't.

A summary list of considerations for decision-making under deeper uncertainty might include:

- *Complexity*: Face up to interconnected, systemic risks; a distant event can disrupt a fragile supply chain.
- *Redundancy*: Don't try to optimize; there will always be wastage.
- *Humility*: Do not think you have thought of everything – there are no 'typical' extreme events.
- *Robustness*: Aim to perform reasonably well over the situations you have thought of.
- *Resilience*: Try to be able to recover quickly, whatever happens.
- *Reversibility*: When there are potentially catastrophic losses, avoid going down a path of no return.
- *Adaptivity*: Have early-warning systems for new issues, and agility to change direction as conditions change.

---

* The fact that this event justified the highest possible level on INES suggests a new scale is needed for nuclear harm. Perhaps it should go up to 11.

- *Openness*: Focus on communication and collaboration, inviting a wide range of perspectives, engage in active deliberation, and avoid taking a single view.
- *Balance*: Don't just focus on the downsides, but think of the potential benefits and harms of any intervention, including possible side effects of being precautionary.

My personal view is that such strategies for dealing with deeper uncertainties may naturally lead to a (possibly temporary) precautionary approach – we don't need a separate principle.

And although this list was intended to apply to organizations and governments, much of it seems relevant to what we have to decide to do every day. We cannot deconstruct every decision and go through the formal process of assessing outcomes, probabilities and values. Instead we need to protect ourselves against the worst that can happen, while taking advantage of the opportunities that arise. To go on adventurous holidays, but with planning, reliable support, and insurance. To take risks, but not be reckless.

•

Summary

- The theoretical basis for rational decision-making in the face of uncertainty requires full specification of options, outcomes, values and probabilities.
- The principle of maximizing expected value does not adequately describe human behaviour.
- Lack of confidence in probabilities generally leads to risk aversion.
- Strategies for decisions become less quantitative as uncertainty grows.
- While we may use imagined narratives in our personal decisions, this seems an inadequate basis for societal decisions.
- The Tolerability of Risk framework places explicit boundaries for intolerable and broadly acceptable risks.
- Uncertainty Factors are used to decide acceptable exposure levels when the appropriate risks are unquantifiable, but these can be excessively cautious.
- The (weak) precautionary principle suggests preventive action without waiting for proof of harm, but is a natural consequence of a resilient strategy for dealing with deep uncertainty.

CHAPTER 16

The Future of Uncertainty

So far we've dealt with some utterly trivial issues, such as the probability of pulling matching socks out of a drawer or of getting a box of double-yolked eggs, and also some serious questions about the risks of cancer and being killed at work. But, as we approach the end of the book, it's time to turn to the big, existential stuff.

What's the probability of you being born? Or for human life to exist? And does it make sense to even ask these questions?

Back in the Introduction, we saw that each of us wouldn't be here were it not for a chain of apparently fortuitous occurrences. And it's now thought that this is not just true for single individuals, but for the whole of humanity. The sequence of events that led to intelligent life developing on Earth can be considered extremely lucky, whether it's the exact distance of Earth from the sun allowing water to exist, the chemical composition of the Earth supporting life, mass extinction of the dinosaurs following an asteroid impact encouraging mammals to develop, and so on.

Even more fundamental, the physical constants underlying the entire universe appear to be finely tuned to support existence, from the precise determinants behind the expansion of the universe, to the window of gravity that allows stars to form, and the ratio of matter to antimatter allowing the universe to coalesce after the Big Bang. Without these 'cosmic coincidences', none of what exists would have developed as it has. It does seem extraordinary that we are here at all.

But these questions are about things that have occurred, and if they hadn't occurred, then we would not be here to ask the question. The *anthropic principle* says it's therefore a bit pointless to even discuss the probability of our existence. It's an extreme example of 'survivorship bias' – books at airports are full of tips about what made successful businesses, but it is impossible to identify factors that increase your chances by only looking at successes – we need to compare them with failures. Similarly, to answer these sorts of existential questions, we would need to consider situations in which we did not exist. Which we don't have access to, unless we perhaps imagine a multiverse in which just one of the possibilities was the Goldilocks zone that allowed us to be here to witness our good fortune.

The problems with trying to answer these questions become even clearer once we accept that all probabilities are judgements expressing personal uncertainty. Who would be the person (or something) assessing these numbers? In some circumstances it can make sense to assess probabilities for events that occurred, such as when someone has won a lottery or had three children with the same birthdays, since it is just about imaginable that we might do this before the event. But nobody would conceivably have thought of me existing before I did, and to assess probabilities about humanity or the universe requires us to think of either an alien species or something outside our universe.

Which doesn't seem to make much practical sense, so that's why I don't have to try to answer these questions. But, sadly, it seems more reasonable to ask the next question.

What's a reasonable probability of humanity ending in the near future?

We know that the world is doomed when the sun expands, although our guaranteed expiry-date is not for a few billion years or so. Within a more foreseeable timescale, there is increasing interest in global catastrophic risks that can threaten the whole of humanity, such as nuclear war, rogue AI, escaped pathogens, asteroid impact, and so on. While we might be able to imagine possible futures in which these existential threats occur, it is challenging to assess probabilities of them occurring.

But that hasn't stopped people trying. Techniques for assessing these extinction probabilities range from purely subjective judgements, such as Astronomer Royal Martin Rees's 'I think the odds are no better than fifty-fifty that our present civilization on Earth will survive to the end of the present century', to a survey of AI experts estimating a rather precise 30.5% probability that an AI catastrophe kills the vast majority of humanity by 2200,[1] to a model-based assessment of a 1 in 2 trillion (million million) chance of a humanity-destroying asteroid strike within a hundred years.[2] Attendees at a Global Catastrophic Risk Conference in 2008, who have a certain interest in the matter, provided a median estimate of 19% risk for extinction by 2100.[3] You may suspect this select group would have an exaggerated sense of threat, but they only estimated a 60% probability for more than 1 million people dying in a natural pandemic by 2100, which seems rather conservative, particularly as this subsequently happened within fifteen years.

While it seems sensible to think seriously about potential threats to humanity, I am not convinced these probabilities are much more than an expression of concern. I would prefer to use the judgements of trained super-forecasters, although I will not be around to score their assessments.

Artificial intelligence

There is a lot of uncertainty about the development and impact of AI. It is certainly going to have an increasing role in all our lives – I used Large Language Models (LLMs) in writing this book, both helping me with coding and research (although I checked and rewrote all their claims). But there is another important issue – how does AI deal with its own uncertainty?

The importance of AI being able to handle uncertainty has been recognized for many years – in fact I contributed to the first 'Uncertainty in AI' conferences way back in the 1980s.[4] Then the major disputes were about whether probabilistic reasoning could be adapted to the complex network structures embodied in what were often termed *expert systems*, and I was part of a group that showed that Bayes' theorem could be used to rigorously propagate uncertainty through chains of reasoning.

Unfortunately, all that work seems to have been lost when it comes to the massive deep-learning networks underlying large-language models, which we know can come up with palpably untrue conclusions delivered with absolute confidence. We've emphasized in this book how people who deliberately under- or overstate uncertainty are promoting disinformation, but at least they have some interest in what is true, even if they want to obscure it. In contrast, large-language models currently have no awareness of whether what they say is true or not: everything is delivered with equal confidence. They can be, to use a technical term, *bullshitters.*[5*]

Bringing uncertainty into deep learning is an active area of research, and uncertainty quantification is an integral part of machine learning in general, and so maybe things will improve and future AI will be able to provide a reliable assessment of its (personal) uncertainty. If these assessments could be shown to be calibrated, this would help enormously in establishing the trustworthiness of AI. I would like my bot to know what it does and doesn't know.

A manifesto for uncertainty

We've covered a lot of material in this book, and congratulations for getting to the end (unless you've jumped to this point). I hope the narrative has seemed coherent, and now it's time to try to take stock, and extract from all the examples and stories some general lessons for improving the handling of uncertainty in the future, both as individuals and as a society.

The first lesson we met is that uncertainty is a personal *relationship* with the world, with an object of attention, a source, an expression and other characteristics. While in some circumstances we can all agree what is going on, in general we have to make our own judgements – we need to *own our uncertainty*, and not only acknowledge it but positively give it a welcome as providing an opportunity for learning and change. This requires honesty and humility, both in self-reflection and in communication with others.

Second, we should endeavour to express our uncertainty through numbers wherever possible. The theory of probability is a remarkable framework with many extraordinary properties, helping us to understand both why coincidences happen so often, and the role of luck in our lives. We can, in principle, claim any probabilities we want, but if we want them to be useful then they need to fit the real world, by being both calibrated and discriminatory. But we should keep in mind that, apart from at the subatomic level, probabilities are not a property of the outside world – they don't actually exist. Although

---

* Philosopher Harry Frankfurt defines 'bullshit' as speech intended to persuade without any regard to truth; while a liar at least acknowledges the truth and tries to hide it, a bullshitter doesn't care if what they say is true or false.

sometimes it can be useful to act as if there are objective 'chances' that determine what happens.

Once we accept that probabilities are essentially judgements, it is natural to take a Bayesian approach to learning from experience, in which our probabilities are updated by accumulating evidence, and Bayes' theorem shows how probability can be used to counter claims that events must have been 'more than a coincidence'. Our brains appear to work in a Bayesian way, with strong prior expectations being updated in the light of observations, but Oliver Cromwell has taught us that we need to keep a small amount of doubt about our understanding of the world so that we can accept and adapt to surprising events.

Science has established ways of acknowledging uncertainty, and formal statistical models are useful, but we should not be deluded into thinking they represent reality. Calculated measures such as P-values and confidence intervals are conditional on all the assumptions of the model being true, which we know is not the case, and so the outputs from statistical packages must always be treated with caution and as rough guides. Claims based on models need sensitivity analysis, recognition of their limitations, and summaries of the quality of the underlying evidence, and preferably assessments of confidence in the whole analysis. Models can be valuable in giving an idea of who, or what, was to blame for bad things that have happened.

Predictions, whether short term or way into the future, can also be based on models of how the world works, and we can assess the quality of the resulting probabilities using scoring rules. This all assumes, of course, that we can list the possible futures, but in more complex circumstances we may be faced with deep uncertainty, in which we acknowledge our understanding is incomplete. Nevertheless we can still try to put our uncertainty into numbers, although this may mean allocating a certain probability to 'something else'.

Extreme and potentially catastrophic events present a particular challenge, although flexible modelling using 'fat-tails' should mean we are less surprised at what turns up, although again it is valuable to have multiple, and very imaginative, views about what may lie in store for us, possibly including some deliberately challenging 'red teams'. Although the UK Ministry of Defence's Red Teaming Handbook argues that it may not be necessary to establish an actual team, but rather that it is vital to have a *red team mindset* that is aware of all the cognitive and behavioural biases that can occur when organizations try to plan the future.[6]

We need to be on our guard against both those who make over-confident claims and, at the other extreme, deliberately try to sow misunderstanding by unduly exaggerating uncertainty. We should expect trustworthy communication, in which conclusions are drawn with humility and uncertainty and are proclaimed with confidence and empathy. But living with uncertainty does not mean being over-cautious – we can take risks without being reckless, while being adaptable and resilient.

These are the personal lessons I have drawn from nearly fifty years working on probability, chance, risk, ignorance and luck. I hope they resonate with you.

We cannot avoid uncertainty. So we need to embrace it, be humbled by it, and even try to enjoy it.

Acknowledgements

My first debt is to Adrian Smith for convincing me, fifty years ago, that probability was a vital and fascinating idea – and that it didn't exist. Since then numerous colleagues have helped me try and get a grip on all the topics in this book, including Philip Dawid, Andrew Gelman, Jerry Toner, Timandra Harkness, Dan Hillman, Miles Hodgkiss, Esther Eidinow, Tim Palmer, David Stainforth, David Flusfeder, David Hand, James Grime, David Tong, Ed Humpherson, Andy Stirling, all the statistical team on the Infected Blood Inquiry, Kevin Mousley at the BBC, and Alex Freeman and everyone at the Winton Centre for Risk and Evidence Communication. And, of course, Ron Biederman for helping me with the story of his remarkable trousers.

I thank Laura Stickney of Penguin for not only commissioning the book, but remaining calm during the kerfuffle when (as usual) we could not agree on a title until the last moment. And credit to Jonathan Pegg for negotiating me a fine deal, Sarah Day for her meticulous editing, and all the staff at Penguin for their excellent support, including Fahad Al-Amoudi, Ruth Pietroni, Julie Woon and Annabel Huxley.

I am indebted to numerous people who read and commented on material, including Maria Skoularidou, Claudia Schneider, Thomas King, Kevin McConway, Alex Freeman, Ken McCallum, Michael Blastland, George Davey-Smith, Stephen Evans, John Kerr, and Vern Farewell – I hope the result is reasonable.

Finally, I must thank Kate Bull for not only her patience in dealing with me, and her vital comments on the text, but also for being a perfect partner and companion, especially when travelling and writing.

While I readily acknowledge the help of Anthropic's *Claude.ai* for coding and research, unfortunately I have to take full responsibility for the inevitable remaining errors and limitations in this book.

Notes

All webpages were accessible in December 2023.

INTRODUCTION

1. J. Toner, *Risk in the Roman World* (Cambridge University Press, 2023).
2. Ipsos MORI, *What Worries the World* (2022), https://www.ipsos.com/en-uk/what-worries-world-december2022. Based on a 'Representative sample of 19,504 adults aged 16–74 in 29 participating countries, 25 Nov.–9 Dec. 2022'.
3. Gallup Inc, *Millennials: The Job-Hopping Generation* (2016), https://www.gallup.com/workplace/231587/millennials-job-hopping-generation.aspx
4. D. Kahneman, *Thinking, Fast and Slow* (Farrar, Straus and Giroux, 2011).
5. S. Žižek, 'Rumsfeld and the Bees', *Guardian,* 28 June 2008.

CHAPTER 1: UNCERTAINTY IS PERSONAL

1. Interview with Esther Eidinow in BBC Radio 4 *Risk Makers*, https://www.bbc.co.uk/programmes/m0002rq8
2. This is adapted from M. Smithson, *Ignorance and Uncertainty: Emerging Paradigms* (Springer, 1989), who uses 'the conscious, metacognitive awareness of ignorance', but the term 'metacognitive' seems redundant.
3. The UK Supreme Court judgement is at *Ivey v Genting Casinos (UK) Ltd (t/a Crockfords) UKSC 67* (2017), http://www.bailii.org/uk/cases/UKSC/2017/67.html
4. A.-R. Blais and E. U. Weber, 'A Domain-Specific Risk-Taking (DOSPERT) scale for adult populations', *Judgment and Decision Making* 1 (2006), 33–47.
5. R. N. Carleton et al., 'Increasingly certain about uncertainty: intolerance of uncertainty across anxiety and depression', *Journal of Anxiety Disorders* 26 (2012), 468–79.

6. M. A. Hillen et al., 'Tolerance of uncertainty: conceptual analysis, integrative model, and implications for healthcare', *Social Science & Medicine* 180 (2017), 62–75.

7. G. Gigerenzer and R. Garcia-Retamero, 'Cassandra's regret: the psychology of not wanting to know', *Psychological Review* 124 (2017), 179–96.

8. Richard Feynman's comments are part of a BBC interview, YouTube, https://www.youtube.com/watch?v=E1RqTP5Unr4

9. The BSE inquiry is at https://webarchive.nationalarchives.gov.uk/ukgwa/20060802142310/http://www.bseinquiry.gov.uk/

10. P. Slovic, 'Perception of risk', *Science* 236 (1987), 280–85.

11. H. P. Lovecraft, *Supernatural Horror in Literature*, https://gutenberg.net.au/ebooks06/0601181h.html

CHAPTER 2: PUTTING UNCERTAINTY INTO NUMBERS

1. P. Wyden, *Bay of Pigs: The Untold Story* (Jonathan Cape, 1979).

2. P. Knapp et al., 'Comparison of two methods of presenting risk information to patients about the side effects of medicines', *Quality and Safety in Health Care* 13 (2004), 176–80.

3. 'Summary of product characteristics. Section 4.8: Undesirable effects', *European Medicines Agency* (2016), https://www.ema.europa.eu/en/documents/presentation/presentation-section-48-undesirableeffects_en.pdf

4. 'MI5 terrorism threat level', MI5 (2010), https://www.mi5.gov.uk/threats-and-advice/terrorism-threat-levels

5. 'The UK National Threat Level has been raised from substantial to SEVERE – meaning an attack is highly likely', Gov.uk, https://www.gov.uk/government/news/uk-terrorism-threat-level-raised-to-severe

6. D. V. Budescu et al., 'The interpretation of IPCC probabilistic statements around the world', *Nature Climate Change* 4 (2014), 508–12.

7. Ibid.

8. D. Irwin and D. Mandel, 'Variants of vague verbiage: intelligence community methods for communicating probability', https://papers.ssrn.com/abstract=3441269

9. Fifth Assessment Report, Summary for Policymakers, Intergovernmental Panel on Climate Change (IPCC) (2014), https://ar5-syr.ipcc.ch/topic_summary.php

10. D. R. Mandel and D. Irwin, 'Facilitating sender–receiver agreement in communicated probabilities: is it best to use words, numbers or both?', *Judgment and Decision Making* 16 (2021), 363–93.

11. Budescu et al., 'The interpretation of IPCC probabilistic statements around the world'.

12. Obama's interview is in the Channel 4 programme *Bin Laden: Shoot to Kill* (2011).

13. J. A. Friedman and R. Zeckhauser, 'Handling and mishandling estimative probability: likelihood, confidence, and the search for Bin Laden', *Intelligence and National Security* 30 (2015), 77–99.

14. 'The death of Osama bin Laden: how the US finally got its man', *Guardian*, 12 Oct. 2012.

15. T. Gneiting et al., 'Probabilistic forecasts, calibration and sharpness', *Journal of the Royal Statistical Society: Series B* 69 (2007), 243–68.

16. Nate Silver's Trump forecast of 28.6% is at https://projects.fivethirtyeight.com/2016-election-forecast/

17. 'Nate Silver's model gives Trump an unusually high chance of winning. Could he be right?', *Vox*, 3 Nov. 2016.

18. R. M. Cooke, 'The aggregation of expert judgment: do good things come to those who weight?', *Risk Analysis* 35 (2015), 12–15.

19. P. E. Tetlock and D. Gardner, *Superforecasting: The Art and Science of Prediction* (McClelland & Stewart, 2015).

20. D. Gardner, *Future Babble* (Penguin, 2012).

21. D. J. Spiegelhalter et al., 'Bayesian approaches to randomized trials', *Journal of the Royal Statistical Society: Series A* 157 (1994), 357–87.

22. N. Dallow et al., 'Better decision making in drug development through adoption of formal prior elicitation', *Pharmaceutical Statistics* 17 (2018), 301–16.

23. Cooke, 'The aggregation of expert judgment'.

24. 'Nulty & Ors v Milton Keynes Borough Council [2013] 1 WLR 1183', para 37, England and Wales Court of Appeal (2013), https://www.casemine.com/judgement/uk/5a8ff70260d03e7f57ea5959

25. Lord Leggatt, 'Some questions of proof and probability', UK Supreme Court, https://www.supremecourt.uk/news/speeches.html#2023

26. M. K. B. Parmar et al., 'The chart trials: Bayesian design and monitoring in practice', *Statistics in Medicine* 13 (1994), 1297–312.

27. M. K. B. Parmar et al., 'Monitoring of large randomised clinical trials: a new approach with Bayesian methods', *Lancet* 358 (2001), 375–81.

CHAPTER 3: TAMING CHANCE WITH PROBABILITY

1. F. N. David, *Games, Gods, and Gambling: A History of Probability and Statistical Ideas* (Dover Publications, 1998).

2. Ibid.

3. G. Cardano, *Liber de ludo aleae* (FrancoAngeli, 2006).

4. 'GCSE Maths Past Papers – Revision Maths', *Edexcel*, at https://revisionmaths.com/gcse-maths/gcse-maths-past-papers/edexcel-gcse-maths-past-papers

5. 'Student protest against "unfair" GCSE maths question goes viral', *Guardian*, 5 June 2015.

6. 'Number of Atoms in the Universe', *Oxford Education Blog*, https://educationblog.oup.com/secondary/maths/numbers-of-atoms-in-the-universe

7. 'Card Shuffling – 52 Factorial', *QI*, at https://www.youtube.com/watch?v=SLIvwtIuC3Y

8. 'Stigler's law of eponymy', Wikipedia.

9. S. M. Stigler, *Casanova's Lottery: The History of a Revolutionary Game of Chance* (University of Chicago Press, 2022).

10. Ibid.

11. 'National Lottery (United Kingdom)', Wikipedia.

12. F. P. Ramsey, 'Truth and probability', McMaster University Archive for the History of Economic Thought; (1926), 156–98, at https://econpapers.repec.org/bookchap/hayhetcha/ramsey1926.htm

13. R. Feynman, 'Probability', *The Feynman Lectures on Physics Vol 1*, Ch. 6, at https://www.feynmanlectures.caltech.edu/I_06.html

14. A. M. Turing, 'The applications of probability to cryptography', www.nationalarchives.gov.uk HW 25/37 (1941–2). A typeset version is at https://arxiv.org/abs/1505.04714

15. B. de Finetti, *Theory of Probability* (Wiley, 1974).

16. 'De Finetti's theorem', Wikipedia.

CHAPTER 4: SURPRISES AND COINCIDENCES

1. 'Cambridge coincidences collection', *Understanding Uncertainty*, https://understandinguncertainty.org/coincidences/

2. 'Ron Biederman's trousers', *Understanding Uncertainty*, https://understandinguncertainty.org/user-submitted-coincidences/ron-biedermans-trousers

3. P. Diaconis and F. Mosteller, 'Methods for studying coincidences', *Journal of the American Statistical Association* 84 (1989), 853–61.

4. 'Army coat hanger', *Understanding Uncertainty*, http://understandinguncertainty.org/user-submitted-coincidences/army-coat-hanger

5. 'Born in the same bed', *Understanding Uncertainty*, http://understandinguncertainty.org/user-submitted-coincidences/born-same-bed

6. 'What are the Odds?', BBC Sounds, https://www.bbc.co.uk/sounds/play/b09v2x58

7. 'Happy birthday to you: couple have 3 children all born on same date', *Daily Mail Online*, 13 Oct. 2010.

8. 'Archive on 4 – Good luck, Professor Spiegelhalter', BBC Sounds, https://www.bbc.co.uk/sounds/play/b09kpmys

9. T. S. Nunnikhoven, 'A birthday problem solution for nonuniform birth frequencies', *American Statistician* 46 (1992), 270–74.

10. 'September 19th is Huntrodds day!', *Understanding Uncertainty*, https://understandinguncertainty.org/september-19th-huntrodds-day

11. O. Flanagan, 'Huntrodds' Day: celebrating coincidence, chance and randomness', *Significance*, 15 Sept. 2014.

12. 'Population estimates by marital status and living arrangements, England and Wales', *Office for National Statistics*, https://www.ons.gov.uk/peoplepopulationandcommunity/populationandmigration/populationestimates/data sets/populationestimatesbymaritalstatusandlivingarrangements

13. 'It's lucky eight for Pagham couple', *Sussex World*, 7 Aug. 2008.

14. R. Sheldrake, 'Morphic resonance and morphic fields: an introduction', https://www.sheldrake.org/research/morphic-resonance/introduction

15. P. Diaconis and F. Mosteller, 'Methods for studying coincidences'.

16. 'To Infinity and beyond', *BBC Horizon 2009–2010*, https://www.bbc.co.uk/programmes/b00qszch

17. A. B. Russell, 'What is the monkey simulator?' (2014), https://github.com/arussell/infinite-monkey-simulator

18. 'Understanding uncertainty: infinite monkey business', *Plus Maths*, https://plus.maths.org/content/infinite-monkey-businesst

19. K. Yates, 'The unexpected maths problem at work during the Women's World Cup', BBC Future, https://www.bbc.com/future/article/20230830-the-unexpected-maths-problem-at-work-during-the-womens-world-cup

20. L. Takács, 'The problem of coincidences', *Archive for History of Exact Sciences* 21 (1980), 229–44.

21. Diaconis and Mosteller, 'Methods for studying coincidences'.

22. D. Spiegelhalter, *The Art of Statistics: Learning from Data* (Penguin, 2019).

23. Every detail of plane crashes can be obtained from https://www.planecrashinfo.com/.

24. D. Spiegelhalter 'Another tragic cluster – but how surprised should we be?', *Understanding Uncertainty*, https://understandinguncertainty.org/another-tragic-cluster-how-surprised-should-we-be

25. 'Statistics and the law', Royal Statistical Society, https://rss.org.uk/membership/rss-groups-and-committees/sections/statistics-law/

26. D. J. Spiegelhalter and H. Riesch, 'Don't know, can't know: embracing deeper uncertainties when analysing risks', *Philosophical Transactions of the Royal Society*, A 369 (2011), 4730–50.

CHAPTER 5: LUCK

1. R. Doll, 'Commentary: the age distribution of cancer and a multistage theory of carcinogenesis', *International Journal of Epidemiology* 33 (2004), 1183–4.

2. '1949 Manchester BEA Douglas DC-3 Accident', Wikipedia.

3. 'Five survivors of spectacular falls', BBC News, 17 June 2013.

4. D. Flusfeder, *Luck: A Personal Account of Fortune, Chance and Risk in Thirteen Investigations* (4th Estate, 2022).

5. 'Archive on 4 – Good luck, Professor Spiegelhalter', BBC Sounds, https://www.bbc.co.uk/sounds/play/b09kpmys

6. 'Edward F. Cantasano', Wikipedia.

7. D. Hadert, 'Lord Howard de Walden', *Guardian*, 12 July 1999.

8. D. K. Nelkin, 'Moral luck', *The Stanford Encyclopedia of Philosophy*, ed. E. N. Zalta and U. Nodelman, https://plato.stanford.edu/archives/spr2023/entries/moral-luck/

9. T. Nagel, *Mortal Questions* (Cambridge University Press, 1979).

10. 'Richard P. Feynman Quote', *A–Z Quotes*, https://www.azquotes.com/quote/1285990

11. 'Early Space Shuttle flights riskier than estimated', *National Public Radio*, 4 March 2011.

12. Unfortunately, the original 2011 report is no longer available on the website for the NASA Space Shuttle Safety and Mission Assurance Office, but the main graphic has been reproduced in D. Spiegelhalter et al., 'Visualizing uncertainty about the future', *Science* 333 (2011), 1393–400 (Supplementary material).

13. 'England's result against India in the third test could hinge on the toss of a coin: I should know . . . I lost 14 in a row!', *Daily Mail Online*, 24 Nov. 2016.

14. 'Derren Brown – 10 Heads in a Row' (2012), YouTube, https://www.youtube.com/watch?v=XzYLHOX50Bc

15. 'Flipping 10 heads in a row: full video' (2011), YouTube, https://www.youtube.com/watch?v=rwvIGNXY21Y

16. 'Builders picking Lotto ball 39 had best chance of winning UK national lottery in 2022', *Guardian*, 27 Dec. 2022.

17. M. J. Mauboussin, *The Success Equation* (Harvard Business Review Press, 2012).

18. 'Football results, statistics & soccer betting odds data', https://www.football-data.co.uk/data.php

19. 'TrueSkill™ ranking system', Microsoft Research, https://www.microsoft.com/enus/research/project/trueskill-ranking-system/

20. E. C. Marshall and D. J. Spiegelhalter, 'Reliability of league tables of in vitro fertilisation clinics: retrospective analysis of live birth rates', *British Medical Journal*, 316 (1998), 1701–4.

21. H. Goldstein and D. J. Spiegelhalter, 'League tables and their limitations: statistical issues in comparisons of institutional performance', *Journal of the Royal Statistical Society: Series A (Statistics in Society)* 159 (1996), 385–409.

22. E. Smith, *Luck, What It Means and Why It Matters* (Bloomsbury, 2012).

23. R. Wiseman, *The Luck Factor* (Arrow, 2004).

24. Detailed outcomes of surgery on children with congenital heart disease are available at https://www.childrensheartsurgery.info/

CHAPTER 6: IT'S ALL A BIT RANDOM

1. A. Lee et al., 'BOADICEA: a comprehensive breast cancer risk prediction model incorporating genetic and nongenetic risk factors', *Genetics in Medicine* 21 (2019), 1708–18.

2. The Cystic Fibrosis Foundation; https://www.cff.org/intro-cf/cf-genetics-basics

3. M. Blastland, *The Hidden Half: How the World Conceals its Secrets* (Atlantic Books, 2019).

4. P. S. Laplace, *A Philosophical Essay on Probabilities* (1814), https://www.gutenberg.org/ebooks/58881

5. D. Garisto, 'The universe is not locally real, and the physics Nobel Prize winners proved it', *Scientific American* (2023).

6. A. Albrecht and D. Phillips, 'Origin of probabilities and their application to the multiverse', *Physical Review D* 90 (2014), 123514.

7. B. B. Brown, 'Some tests on the randomness of a million digits', RAND Corporation (1948), https://www.rand.org/pubs/papers/P44.html

8. Rand Corporation, *A Million Random Digits with 100,000 Normal Deviates* (Rand Corporation, 2001).

9. G. W. Brown, 'History of RAND's random digits: summary', RAND Corporation (1949), https://www.rand.org/pubs/papers/P113.html

10. 'Tails you win: the science of chance', BBC Four, https://www.bbc.co.uk/programmes/p00yh2rc

11. P. Diaconis et al., 'Dynamical bias in the coin toss', *SIAM Review* 49 (2007), 211–35.

12. E. Paparistodemou et al., 'The interplay between fairness and randomness in a spatial computer game', *International Journal of Computing and Machine Learning* 13 (2008), 89–110.

13. 'U.S. makes mistake on Visa lottery, must redraw', Reuters, 13 May 2011.

14. 'Lottery draft – 1969, CBS News', YouTube, http://www.youtube.com/watch?v=-p5X1FjyD_g

15. 'UK national lotto winning numbers', http://lottery.merseyworld.com/Winning_index.html

16. John Haigh showed the chi-squared statistic must be increased by a factor 48/43, before being compared to a null distribution with 48 degrees of freedom. The resulting P-values for the four distributions are 0.97, 0.34, 0.12 and 0.21, showing good compatibility with a uniform distribution. J. Haigh, 'The statistics of the National Lottery', *Journal of the Royal Statistical Society: Series A* 160 (1997), 187–206.

17. 'How to win lotto: Beat Lottery', *BeatLottery.co.uk*, https://www.beatlottery.co.uk/lotto/how-to-win

18. 'Stephanie Shirley career story: the importance of being ERNIE', *Significance* 3 (2006), 33–6.

19. A. L. Mishara, 'Klaus Conrad (1905–1961): delusional mood, psychosis, and beginning schizophrenia', *Schizophrenia Bulletin* 36 (2010), 9–13.

20. B. Cohen, 'Spotify made its shuffle feature less random so that it would actually feel more random to listeners – here's why', *Business Insider* (2020), https://www.businessinsider.com/spotify-made-shuffle-feature-less-random-to-actually-feel-random-2020-3

21. I. Palacios-Huerta, 'Professionals play Minimax', *Review of Economic Studies* 70 (2003), 395–415.

22. N. M. Laird, 'A conversation with F. N. David', *Statistical Science* 4 (1989), 235–46.

CHAPTER 7: BEING BAYESIAN

1. T. Bayes, 'An essay towards solving a problem in the doctrine of chances', *Philosophical Transactions* 53 (1763), 370–418.

2. E. O'Dwyer, 'Facial recognition cameras set to scan crowds at King's coronation as 11,500 police deployed', *inews.co.uk*, 3 May 2023.

3. 'Live facial recognition', *College of Policing* (2022), https://www.college.police.uk/app/live-facial-recognition/live-facial-recognition

4 'Met police to deploy facial recognition cameras', BBC News, 24 Jan. 2020.

5. S. Coble, 'London police adopt facial recognition technology as Europe considers five-year ban', *Infosecurity Magazine* (2020), https://www.infosecurity-magazine.com/news/the-met-adopt-facial-recognition/

6. 'Alan Turing papers on code breaking released by GCHQ', BBC News, 19 April 2012.

7. D. Spiegelhalter, *The Art of Statistics: Learning from Data* (Penguin, 2019).

8. 'The influence of ULTRA in the Second World War', https://www.cix.co.uk/~klockstone/hinsley.htm

9. T. Carlyle, *Oliver Cromwell's Letters and Speeches: with elucidations* (Scribner, Welford and Co., 1871). Available from: http://www.gasl.org/refbib/Carlyle__Cromwell.pdf

10. R. Bain, 'Are our brains Bayesian?', *Significance* 13 (2016), 14–19.

CHAPTER 8: SCIENCE AND UNCERTAINTY

1. 'GUM: guide to the expression of uncertainty in measurement', BIPM (2008), https://www.bipm
.org/en/committees/jc/jcgm/publications

2. B. N. Taylor, 'Guidelines for evaluating and expressing the uncertainty of NIST measurement
results', United States: Commerce Department: National Institute of Standards and Technology
(NIST), National Bureau of Standards (U.S.) (1993), http://dx.doi.org/10.6028/NIST.TN.1297

3. M. Henrion and B. Fischhoff, 'Assessing uncertainty in physical constants', *American Journal of
Physics* 54 (1986), 791–8.

4. Ibid.

5. A. D. Franklin, 'Millikan's published and unpublished data on oil drops', *Historical Studies in the
Physical Sciences* 11 (1981), 185–201.

6. The RECOVERY Collaborative Group, 'Dexamethasone in hospitalized patients with Covid-19',
New England Journal of Medicine 384 (2021), 693–704.

7. E. Thompson, *Escape from Model Land* (Basic Books, 2022).

8. G. E. P. Box, 'Science and statistics', *Journal of the American Statistical Association* 71 (1976), 791–9.

9. R. L. Wasserstein and N. A Lazar, 'The ASA statement on p-values: context, process, and purpose',
American Statistician 70 (2016), 129–33.

10. Ibid.

11. S. Greenland et al., 'To curb research misreporting, replace significance and confidence by compatibility', *Preventive Medicine* 164 (2022), 107127.

12. 'COVID treatment developed in the NHS saves a million lives', NHS England (2021).

13. R. M. Turner et al., 'Routine antenatal anti-D prophylaxis in women who are Rh(D) negative: meta-
analyses adjusted for differences in study design and quality, *PLOS ONE* (2012), e30711.

14. J. Park et al., 'Combining models to generate a consensus effective reproduction number R for
the COVID-19 epidemic status in England', *medRxiv* (2023), https://www.medrxiv.org/content/10
.1101/2023.02.27.23286501v1

15. 'SPI-M-O: consensus statement on COVID-19', Gov.uk, 15 Oct. 2020.

16. T. Maishman et al., 'Statistical methods used to combine the effective reproduction number, R(t),
and other related measures of COVID-19 in the UK', *Statistical Methods in Medical Research* 31
(2022).

17. A. Oza, 'Reproducibility trial: 246 biologists get different results from same data sets', *Nature*, 12
Oct. 2023.

18. D. A. van Dyk, 'The role of statistics in the discovery of a Higgs boson', *Annual Review of Statistics
and Its Applications* 1 (2014), 41–59.

19. 'New results indicate that new particle is a Higgs boson', CERN, 14 March 2013.

20. S. Stepanyan et al., 'Observation of an exotic S = +1 baryon in exclusive photoproduction from the
deuteron', *Physical Review Letters* 91 (2003), 252001.

21. van Dyk, 'The role of statistics in the discovery of a Higgs Boson'.

22. 'Faster than light particles found, claim scientists', *Guardian*, 22 Sept. 2011.

23. https://statmodeling.stat.columbia.edu/2024/03/27/bayesian-inference-with-informative-priors
-is-not-inherently-subjective/

CHAPTER 9: HOW MUCH CONFIDENCE DO WE HAVE IN OUR ANALYSIS?

1. Ministry of Defence, 'Joint doctrine publication 2-00, intelligence, counter-intelligence
and security support to joint operations' (2023), https://assets.publishing.service.gov.uk/
media/653a4b0780884d0013f71bb0/JDP_2_00_Ed_4_web.pdf

2. 'Assessing Russian activities and intentions in recent U.S. elections', Intelligence Committee, 6 Jan. 2019, https://www.intelligence.senate.gov/publications/assessing-russian-activities-and-intentions-recent-us-elections

3. J. A. Friedman and R. Zeckhauser, 'Handling and mishandling estimative probability: likelihood, confidence, and the search for Bin Laden', *Intelligence and National Security* 30 (2015), 77–99.

4 D. Irwin and D. R. Mandel, 'Communicating uncertainty in national security intelligence: expert and nonexpert interpretations of and preferences for verbal and numeric formats', *Risk Analysis* 43 (2023), 943–57.

5 'Assessing Russian activities and intentions in recent U.S. elections'.

6. 'Contaminated blood', UK Parliament, 11 July 2017, https://hansard.parliament.uk/commons/2017-0711/debates/E647265A-4A8A-4D87-95A2-66A91E3A37D6/ContaminatedBlood

7. J. M. Micallef et al., 'Spontaneous viral clearance following acute hepatitis C infection: a systematic review of longitudinal studies', *Journal of Viral Hepatitis* 13 (2006), 34–41.

8. 'Inquiry publishes report by the Statistics Expert Group', Infected Blood Inquiry, 15 Sept. 2022.

9. IPCC Cross-Working Group Meeting on Consistent Treatment of Uncertainties, 'Guidance note for lead authors of the IPCC Fifth Assessment Report on consistent treatment of uncertainties', IPCC, 2010, http://www.ipcc-wg2.gov/meetings/CGCs/Uncertainties-GN_IPCCbrochure_lo.pdf

10. IPCC AR6 Working Group 1, 'Summary for policymakers', IPCC, 2022, https://www.ipcc.ch/report/ar6/wg1/chapter/summary-for-policymakers/

11. A. Kause et al., 'Confidence levels and likelihood terms in IPCC reports: a survey of experts from different scientific disciplines', *Climatic Change* 173 (2022).

12. 'What is GRADE?', *BMJ Best Practice*, https://bestpractice.bmj.com/info/toolkit/learn-ebm/what-is-grade/

13. H. Balshem et al., 'GRADE guidelines: 3. Rating the quality of evidence', *Journal Clinical Epidemiology* 64 (2011), 401–6.

14. Ibid.

15. 'Non-pharmaceutical interventions (NPIs) table', Gov.uk, 21 Sept. 2020, https://www.gov.uk/government/publications/npis-table-17-september-2020/non-pharmaceutical-interventions-npis-table-21-september-2020

16. 'Teaching & learning toolkit', Education Endowment Foundation, 12 May 2016, https://educationendowmentfoundation.org.uk/evidence/teaching-learning-toolkit

17. 'Official statistics in development', Office for Statistics Regulation, https://osr.statisticsauthority.gov.uk/policies/official-statistics-policies/official-statistics-in-development/

CHAPTER 10: WHAT, OR WHO, IS TO BLAME? CAUSALITY, CLIMATE AND CRIME

1. C. J. Ferguson, 'The good, the bad and the ugly: a meta-analytic review of positive and negative effects of violent video games', *Psychiatric Quarterly* 78 (2007), 309–16.

2. 'Can the cat give you cancer? Parasite in their bellies linked with brain tumours', *Daily Mail Online*, 27 July 2011.

3. D. Grady et al., 'Hormone therapy to prevent disease and prolong life in postmenopausal women' *Annals of Internal Medicine* 117 (1992), 1016–37.

4. J. E. Manson et al., 'The Women's Health Initiative hormone therapy trials: update and overview of health outcomes during the intervention and post-stopping phases', *Journal of the American Medical Association* 310 (2013), 1353–68.

5. H. N. Hodis and W. J. Mack, 'Menopausal hormone replacement therapy and reduction of all-cause mortality and cardiovascular disease: it's about time and timing', *Cancer Journal*, 28 (2022), 208–23.

6. H. S. Hansen et al., 'The fraction of lung cancer attributable to smoking in the Norwegian Women and Cancer (NOWAC) Study', *British Journal of Cancer* 124 (2021), 658–62.

7. 'Bacon, ham and sausages have the same cancer risk as cigarettes, warn experts', *Daily Record*, 23 Oct. 2015.

8. J. M. Samet et al., 'The IARC *Monographs*: updated procedures for modern and transparent evidence synthesis in cancer hazard identification', *Journal of the National Cancer Institute* 112 (2019), 30–37.

9. 'Aspartame sweetener to be declared possible cancer risk by WHO, say reports', *Guardian*, 29 June 2023.

10. 'Quantifying uncertainty in causal analysis', US Environmental Protection Agency (2016), https://www.epa.gov/caddis-vol1/quantifying-uncertainty-causal-analysis

11. 'IPCC AR6 Working Group 1: Summary for policymakers', IPCC, https://www.ipcc.ch/report/ar6/wg1/chapter/summary-for-policymakers/

12. Ibid.

13. 'Attributing extreme weather to climate change', Met Office, https://www.metoffice.gov.uk/research/climate/understanding-climate/attributing-extreme-weather-to-climate-change

14. G. Schmidt, 'Climate models can't explain 2023's huge heat anomaly – we could be in uncharted territory', *Nature*, 19 March 2024.

15. F. Guterl et al., 'How global warming is turbocharging monster storms', *Newsweek*, 5 Sept. 2018.

16. K. A. Reed et al., 'Forecasted attribution of the human influence on Hurricane Florence', *Science Advances* 6 (2020).

17. S.-K. Min et al., 'Anthropogenic contribution to the 2017 earliest summer onset in South Korea', *Bulletin of the American Meteorological Society* 100 (2019), S73–7.

18. A. Hannart and P. Naveau, 'Probabilities of causation of climate changes', *Journal of Climate* 31 (2018), 5507–24.

19. 'Adverse drug reaction probability scale (Naranjo) in drug-induced liver injury', *LiverTox: Clinical and Research Information on Drug-Induced Liver Injury*, National Institute of Diabetes and Digestive and Kidney Diseases (2012), http://www.ncbi.nlm.nih.gov/books/NBK548069/

20. 'Novartis Grimsby Ltd v Cookson', England and Wales Court of Appeal, EWCA Civ 1261 (2007).

21. 'FAQs: probability of causation', Centers for Disease Control and Prevention, https://www.cdc.gov/niosh/ocas/faqspoc.html

22. A. Broadbent, 'Epidemiological evidence in proof of specific causation', *Legal Theory* 17 (2011), 237–78.

23. *Reference Manual on Scientific Evidence: Third Edition* (National Academies Press, 2011), http://www.nap.edu/catalog/13163

24. 'DNA-17 Profiling', Crown Prosecution Service, https://www.cps.gov.uk/legal-guidance/dna-17-profiling

25. 'Guideline for evaluative reporting in forensic science', ENFSI (2016), https://enfsi.eu/docfile/enfsiguideline-for-evaluative-reporting-in-forensic-science/

26. 'Science and the law', Royal Society, https://royalsociety.org/about-us/programmes/science-and-law/

27. 'R v Sally Clark', England and Wales Court of Appeal, EWCA Crim 1020 (2003).

28. P. Dawid, 'Statistics on trial', *Significance* 2 (2005), 6–8

29. Ibid.

30. 'R v Sally Clark', England and Wales Court of Appeal, EWCA Crim 1020 (2003).

31. 'R v Adams', Wikipedia.

32. L. H. Tribe, 'Trial by mathematics: precision and ritual in the legal process', *Harvard Law Review* 84 (1971), 1329–93.

33. Lord Leggatt, 'Some questions of proof and probability', UK Supreme Court, https://www
.supremecourt.uk/news/speeches.html#2023

34. J. M. Keynes, *Treatise on Probability* (Macmillan, 1921).

35. Lord Leggatt, 'Some questions of proof and probability'.

36. N. Nic Daéid et al., 'The use of statistics in legal proceedings: a primer for the courts', Royal Society (2020).

CHAPTER 11: PREDICTING THE FUTURE

1. J. K. Rowling, *Harry Potter and the Prisoner of Azkaban* (Bloomsbury, 1999).

2. https://improbability-principle.com/

3. S. D. Snobelen, 'Statement on the date 2060', https://isaac-newton.org/statement-on-the-date-2060/

4. 'More or Less – 22/05/2009', BBC Sounds, https://www.bbc.co.uk/sounds/play/bookfsgg

5. D. J. Spiegelhalter, 'The professor's premiership probabilities', 22 May 2009, http://news.bbc.co
.uk/1/hi/programmes/more_or_less/8062277.stm

6. 'Lawro's predictions', BBC, 24 May 2009, http://news.bbc.co.uk/sport1/hi/football/8048360.stm

7. D. J. Spiegelhalter and Y-L. Ng, 'One match to go!', *Significance* 6 (2009), 151–3.

8. T. Palmer, *The Primacy of Doubt: From Climate Change to Quantum Physics, How the Science of Uncertainty Can Help Predict and Understand our Chaotic World* (Oxford University Press, 2022).

9. Skill scores of forecasts of weather parameters by TIGGE centres, ECMWF, https://charts
.ecmwf.int/products/plwww_3m_ens_tigge_wp_mean?area=Europe&parameter=24h%20
precipitation&sco re=Brier%20skill%20score

10. 'GraphCast: AI model for faster and more accurate global weather forecasting', Google DeepMind, 14 Nov. 2023.

11. 'Inflation report – May 2018', *Bank of England*, https://www.bankofengland.co.uk/inflation
-report/2018/may-2018

12. J. Mitchell and M. Weale, 'Forecasting with unknown unknowns: censoring and fat tails on the Bank of England's Monetary Policy Committee', EMF Research Papers (2019), https://ideas.repec
.org//p/wrk/wrkemf/27.html

13. K. Wijndaele et al., 'Television viewing time independently predicts all-cause and cardiovascular mortality: the EPIC Norfolk Study', *International Journal of Epidemiology* 40 (2011), 150–59.

14. A. Sud et al., 'Realistic expectations are key to realising the benefits of polygenic scores', *British Medical Journal* 380.(2023), e073149.

15. 'Hancock criticised over DNA test "over-reaction"', BBC News, 21 March 2019.

16. 'Predict prostate', https://prostate.predict.nhs.uk/

17. 'Life expectancy for local areas in England, Northern Ireland and Wales: between 2001 to 2003 and 2020 to 2018', Office for National Statistics, 23 Sept. 2021, https://www.ons.gov.uk/
peoplepopulationandcommunity/healthandsocialcare/healthandlifeexpectancies/bulletins/
lifeexpectancyforlocalareasoftheuk/between2001to2003and2018to2020

18. 'Past and projected period and cohort life tables: 2020-based, UK 1981 to 2070', Office for National Statistics, 12 Jan. 2022, https://www.ons.gov.uk/peoplepopulationandcommunity/births
deathsandmarriages/lifeexpectancies/bulletins/pastandprojecteddatafromtheperiodandcohort
lifetables/2020baseduk1981to2070

19. 'Mortality improvements and CMI_2021: frequently asked questions (FAQs)', Institute and Faculty of Actuaries, https://www.actuaries.org.uk/mortality-improvements-and-cmi-2021-frequently
-asked-questions-faqs

20. 'Climate change 2021: the physical science basis. Working Group I Contribution to the IPCC Sixth Assessment Report, Chapter 4', IPCC (2021), https://www.ipcc.ch/report/ar6/wg1/chapter/chapter-4/

21. D. Stainforth, 'The big idea: can we predict the climate of the future?', *Guardian*, 22 Oct. 2023.
22. See for example 'Can policymakers trust forecasters? Experts, modelers, and forecasters try to predict events, but which of them are most reliable?', Institute for Progress (IFP), https://ifp.org/can-policymakers-trust-forecasters/

CHAPTER 12: RISK, FAILURE AND DISASTER

1. J. E. Heffernan and J. A. Tawn, 'An extreme value analysis for the investigation into the sinking of the M. V. *Derbyshire*', *Journal of the Royal Statistical Society Series C: Applied Statistics* 52 (2003), 337–54.
2. F. Liljeros et al., 'The web of human sexual contacts', *Nature* 411 (2001), 907–8.
3. A. Clauset and R. Woodard, 'Estimating the historical and future probabilities of large terrorist events', *Annals of Applied Statistics* 7 (2013), 1838–65.
4. E. Frederick, 'Predicting Three Mile Island', *MIT Technology Review*, 24 April 2019.
5. T. R. Wellock, 'A figure of merit: quantifying the probability of a nuclear reactor accident', *Technological Culture* 58 (2017), 678–721.
6 E. Marsden, 'Farmer's diagram, or F-N curve: representing society's degree of catastrophe aversion', *Risk Engineering*, 22 July 2022, https://risk-engineering.org/concept/Farmer-diagram
7. 'Reactor safety study: an assessment of accident risks in U.S. commercial nuclear power plants. Report NoWASH-1400-MR', Nuclear Regulatory Commission (Washington, DC, 1975), https://www.osti.gov/biblio/7134131
8. 'Flood risk ten times higher in many places over the world within 30 years', Deltares, 23 March 2023, https://www.deltares.nl/en/news/flood-risk-ten-times-higher-in-many-places-over-the-world-within-30-years
9 T. H. J. Hermans et al., 'The timing of decreasing coastal flood protection due to sea-level rise', *Nature Climate Change* 13 (2023), 359–66.
10. 'Your long term flood risk assessment', Gov.uk, https://check-long-term-flood-risk.service.gov.uk/risk
11. 'National risk register 2023', Gov.uk, https://www.gov.uk/government/publications/national-risk-register2023
12. H. Sutherland et al., 'How people understand risk matrices, and how matrix design can improve their use: findings from randomized controlled studies', *Risk Analysis* 42 (2021), 1023–41.
13. The House of Lords Science and Technology Select Committee report, https://publications.parliament.uk/pa/cm201011/cmselect/cmsctech/498/49808.htm
14. World Economic Forum, Global Risks Report 2023. Available from: https://www.weforum.org/publications/global-risks-report-2023/
15 H. W. J. Rittel and M. M. Webber, 'Dilemmas in a general theory of planning', *Policy Science* 4 (1973), 155– 69.
16. J. Wolff, 'Risk, fear, blame, shame and the regulation of public safety', *Economics and Philosophy* 22 (2006), 409–27.

CHAPTER 13: DEEP UNCERTAINTY

1. N. Taleb, *The Black Swan: The Impact of the Highly Improbable* (Random House, 2007).
2. J. Derbyshire, 'Answers to questions on uncertainty in geography: old lessons and new scenario tools', *Environment Planning A: Economy and Space* 52 (2020), 710–27.
3. A. Stirling, 'Keep it complex', *Nature* 468 (2010), 1029–31.
4. J. M. Keynes, 'The General Theory of Employment', *Quarterly Journal of Economics* 51 (1937), 209–23, at 213–14.

5. F. Knight, *Risk, Uncertainty and Profit* (1921), http://www.econlib.org/library/Knight/knRUP.html

6. R. M. Cooke, 'Deep and shallow uncertainty in messaging climate change', *Safety, Reliability and Risk Analysis* (CRC Press, 2013), https://papers.ssrn.com/abstract=2432227

7. D. Benford et al., 'The principles and methods behind EFSA's guidance on uncertainty analysis in scientific assessment', *EFSA Journal* 16 (2018), e05122.

8. R. Flage and T. Aven, 'Expressing and communicating uncertainty in relation to quantitative risk analysis (QRA)', *Reliability and Risk Analysis Theory Applications* 2 (2009), 9–18.

9. J. Kay and M. King, *Radical Uncertainty* (Bridge Street Press, 2020).

10. O. A. Lindaas and K. A. Pettersen, 'Risk analysis and black swans: two strategies for de-blackening', *Journal of Risk Research* 19 (2016), 1231–45.

11. Derbyshire, 'Answers to questions on uncertainty in geography', 710–27.

12. 'Emissions scenarios', IPCC, https://archive.ipcc.ch/ipccreports/sres/emission/index.php?idp=3

13. 'Stories from tomorrow: exploring new technology through useful fiction', Gov.uk, https://www.gov.uk/government/publications/stories-from-the-future-exploring-new-technology-through-useful-fiction/stories-from-tomorrow-exploring-new-technology-through-useful-fiction

CHAPTER 14: COMMUNICATING UNCERTAINTY AND RISK

1. 'Full text of Dick Cheney's Speech', *Guardian*, 27 Aug. 2002.

2. 'September Dossier', Wikipedia (2023).

3. Review of Intelligence on Weapons of Mass Destruction, http://www.butlerreview.org.uk/

4. Report of the Select Committee on Intelligence on prewar intelligence assessments about post-war Iraq together with additional and minority views. Library of Congress, https://www.loc.gov/item/2008354011/

5. Merchants of Doubt, https://www.merchantsofdoubt.org/

6. *Agnotology: The Making and Unmaking of Ignorance* (Stanford University Press, 2008).

7. O. O'Neill, 'Linking trust to trustworthiness', *International Journal of Philosophical Studies* 26(2) (2018), 293–300.

8. B. Fischhoff and A. L. Davis, 'Communicating scientific uncertainty', *Proceedings of the National Academy Sciences* 111 (Supplement 4), 16 Sep. 2014, 13664–71.

9. J. Champkin, 'Lord Krebs', *Significance* 10 (2013), 23–9.

10. B. C. Young et al., 'Daily testing for contacts of individuals with SARS-CoV-2 infection and attendance and SARS-CoV-2 transmission in English secondary schools and colleges: an open-label, cluster-randomised trial', *Lancet* 398 (2021), 1217–29.

11. J. W. Tukey, 'The future of data analysis', *Annals of Mathematical Statistics* 33 (1962), 1–67, at 13–14.

12. S. Teal and A. Edelman, 'Contraception selection, effectiveness, and adverse effects: a review', *Journal of the American Medical Association* 326 (2021), 2507–18.

13. 'What does probability of precipitation mean?', NOAA's National Weather Service, https://www.weather.gov/lmk/pops

14. M. Galesic and R. Garcia-Retamero, 'Statistical numeracy for health: a cross-cultural comparison with probabilistic national samples', *Archives of Internal Medicine*, 170 (2010), 462–8.

15. D. Bourdin and R. Vetschera, 'Factors influencing the ratio bias', *EURO Journal on Decision Processes* 6 (2018), 321–42.

16. 'Binge watching can actually kill you, says new study', *Independent*, 25 July 2016.

17. T. Shirakawa et al., 'Watching television and risk of mortality from pulmonary embolism among Japanese men and women', *Circulation* 134 (2016), 355–7.

18. S. S. Hall, 'Scientists on trial: at fault?', *Nature News* 477 (2011), 264–9.

19. F. P. Polack et al., 'Safety and efficacy of the BNT162b2 mRNA Covid-19 vaccine', *New England Journal of Medicine* 383 (2021), 2603–15.

20. 'Micromort', Wikipedia.

21. L. Padilla et al., 'Uncertainty visualization', in W. Piegorsch et al. (eds.), *Computational Statistics in Data Science* (Wiley, 2022), 405–21.

22. P. K. J. Han et al., 'Communication of uncertainty regarding individualized cancer risk estimates', *Medical Decision Making* 31 (2011), 354–66.

23. C. R. Schneider et al., 'The effects of quality of evidence communication on perception of public health information about COVID-19: two randomised controlled trials', *PLoS One* 16 (2021), e0259048.

24. A. M. van der Bles et al., 'The effects of communicating uncertainty on public trust in facts and numbers', *Proceedings of the National Academy Sciences* 117 (2020), 7672–83.

25. M. Blastland et al., 'Five rules for evidence communication', *Nature* 587 (2020), 362–4.

26. E. Humpherson, 'Uncertainty about official statistics', *Journal of Risk Research* (2024), DOI: 10.1080/13669877.2024.2360920.

27. 'RESIST 2 counter disinformation toolkit', Government Communication Service, https://gcs .civilservice.gov.uk/publications/resist-2-counter-disinformation-toolkit/

28. J. R. Kerr et al., 'Transparent communication of evidence does not undermine public trust in evidence, *PNAS Nexus* 1 (2022), pgac280.

CHAPTER 15: MAKING DECISIONS AND MANAGING RISKS

1. L. J. Savage, *The Foundations of Statistics* (Dover, 1972).

2. A. Tversky and D. Kahneman, 'Advances in prospect theory: cumulative representation of uncertainty', *Journal of Risk and Uncertainty* 5 (1992), 297–323.

3. K. Ruggeri et al., 'Replicating patterns of prospect theory for decision under risk', *Nature Human Behaviour* 4 (2020), 622–33.

4. 'Daniel Kahneman – dyads, and other mysteries', https://josephnoelwalker.com/143-daniel-kahneman/

5. D. Ellsberg, 'Risk, ambiguity and the savage axioms', *Quarterly Journal of Economics* 75 (1961), 643–69.

6. 'Decision support tools', NHS England, https://www.england.nhs.uk/personalisedcare/shared -decision-making/decision-support-tools/

7. G. Gigerenzer and D. G. Goldstein, 'Reasoning the fast and frugal way: models of bounded rationality', *Psychological Review* 103 (1996), 650–69.

8. See for example https://gobraithwaite.com/thinking/how-daniel-kahneman-learned-the-value-of -stories-for-thinking-fast-and-slow/

9. S. G. B. Johnson et al., 'Conviction narrative theory: a theory of choice under radical uncertainty', *Behavioural and Brain Sciences* 30 (2022), e82.

10. 'The Green Book', Gov.uk (2022), https://www.gov.uk/government/publications/the-green-book -appraisal-and-evaluation-in-central-government/the-green-book-2020.

11. 'TPWS – the once and future safety system', *Modern Railways* 25 Sept. 2019.

12. 'Woodland natural capital accounts', Office for National Statistics, https://www.ons.gov.uk/ economy/environmentalaccounts/bulletins/woodlandnaturalcapitalaccountsuk/2022

13. B. Flyvbjerg, 'Top ten behavioral biases in project management: an overview', *Project Management Journal* 52 (2021).

14. 'Valuation – EQ-5D', EuroQol, https://euroqol.org/eq-5d-instruments/eq-5d-5l-about/valuation -standard-value-sets/

15. Health and Safety Executive, *Reducing Risks, Protecting People: HSE's Decision-making Process* (2011), http://www.hse.gov.uk/risk/theory/r2p2.htm

16. 'Flixborough (Nypro UK) Explosion 1st June 1974', Health and Safety Executive, https://www.hse.gov.uk/comah/sragtech/caseflixboroug74.htm

17. 'Union Carbide India Ltd, Bhopal, India. 3rd December 1984', Health and Safety Executive, https://www.hse.gov.uk/comah/sragtech/caseuncarbide84.htm

18. 'Browned toast and crispy roast potatoes "a potential cancer risk"', *Telegraph*, 22 Jan. 2017.

19. D. J. Spiegelhalter, 'How dangerous is burnt toast?' (2017), https://medium.com/wintoncentre/how-dangerous-is-burnt-toast-c5e237873097

20. D. J. Spiegelhalter, 'Coffee and cancer: what Starbucks might have argued' (2018), https://medium.com/wintoncentre/coffee-and-cancer-what-starbucks-might-have-argued-2f20aa4a9fed

21. 'Proposed OEHHA regulation clarifies that cancer warnings are not required for coffee under proposition 65', OEHHA, 15 June 2018.

22. R. Poole et al., 'Coffee consumption and health: umbrella review of meta-analyses of multiple health outcomes', *British Medical Journal* 359 (2017), j5024.

23. O. V Martin et al., 'Dispelling urban myths about default uncertainty factors in chemical risk assessment – sufficient protection against mixture effects?', *Environmental Health* 12 (2013), 53.

24. D. A. Dankovic et al., 'The scientific basis of uncertainty factors used in setting occupational exposure limits', *Journal of Occupational and Environmental Hygiene* 12 (2015), S55–68.

25. J. Peto et al., 'The European mesothelioma epidemic', *British Journal of Cancer* 79 (1999), 666–72.

26. P. E. Tetlock et al., 'False dichotomy alert: improving subjective-probability estimates vs. raising awareness of systemic risk', *International Journal of Forecasting* 39 (2023), 1021–5.

27. D. Spiegelhalter, 'Fear and numbers in Fukushima', *Significance* 8 (2011), 100–103.

28. A. Hasegawa et al., 'Health effects of radiation and other health problems in the aftermath of nuclear accidents, with an emphasis on Fukushima', *Lancet* 386 (2015), 479–88.

CHAPTER 16: THE FUTURE OF UNCERTAINTY

1. 'Treaty on artificial intelligence safety and cooperation', TAISC.org, https://taisc.org

2. J.-M. Salotti, 'Humanity extinction by asteroid impact', *Futures* 138 (2022), 102933.

3. A. Sandberg and N. Bostrom, 'Global catastrophic risks survey', Technical report 2008-1, Future Humanity Institute, University of Oxford, 2008.

4. D. J. Spiegelhalter, 'Probabilistic reasoning in predictive expert systems', in L. N. Kanal and J. Lemmer (eds.), *Uncertainty in Artificial Intelligence* (North-Holland, 1986), pp. 47–68.

5. H. G. Frankfurt, *On Bullshit* (Princeton University Press, 2005).

6. Ministry of Defence, *Red Teaming Handbook*, Gov.uk, https://www.gov.uk/government/publications/a-guide-to-red-teaming

Glossary

Absolute risk: the proportion of people in a defined group who experience an event of interest within a specified period of time.

Aleatory uncertainty: unavoidable unpredictability about the future, also known as chance, randomness, luck, and so on.

Attributable fraction: If RR is the *relative risk*, then for every case in the unexposed group, we expect RR in the exposed group. So, of the RR exposed cases, we would expect 1 of them to have occurred in the absence of exposure. So the attributable or excess fraction is $AF = (RR - 1)/R = 1 - 1/RR$.

Bayes factor: the relative support given by a set of data for two alternative hypotheses. For hypotheses H_0 and H_1, and data x, the ratio $\Pr(x|H_0)/\Pr(x|H_1)$.

Bayes' theorem: a rule of probability that shows how evidence B updates prior beliefs of a proposition A to produce posterior beliefs $\Pr(A|B)$, through the formula $\Pr(A|B) = \dfrac{\Pr(B|A) \times \Pr(A)}{\Pr(B)}$. This is easily proved: since $\Pr(B \text{ and } A) = \Pr(A \text{ and } B)$, the multiplication rule of probability means that $\Pr(A|B)\Pr(B) = \Pr(B|A)\Pr(A)$, and dividing each side by $\Pr(B)$ gives the theorem.

Bayesian: the approach to statistical inference in which probability is used not only for aleatory uncertainty but also for epistemic uncertainty about unknown facts or quantities. Bayes' theorem is then used to revise these beliefs in the light of new evidence.

Bernoulli trial: If X is a random variable which takes on the value 1 with probability p, and 0 with probability $1 - p$, it is known as a Bernoulli trial. X has a Bernoulli distribution with mean p and variance $p(1 - p)$.

Beta distribution: If p is an unknown probability of 'success' in a Bernoulli trial, then in a Bayesian analysis it may be given a beta prior distribution, denoted Beta[a,b], meaning the probability density is $f(p) =$

$\frac{\Gamma(a + b)}{\Gamma(a)\Gamma(b)} p^{a-1}(1 - p)^{b-1}; 0 \leq p \leq 1$. So a Beta[$a,b$] distribution is uniform

on [0,1]. Suppose the sampling distribution is Binomial[p,n], and the observed number of 'successes' is r. Then Bayes' theorem says that the posterior distribution $\propto$ likelihood $\times$ prior, which means the posterior $\propto$ $p^r(1 - p)^{n-r} p^{a-1}(1 - p)^{b-1} = p^{r+a-1}(1 - p)^{n-r+b-1}$, so the posterior density $f(p|r,n)$ is Beta[$r + a, n - r + b$].

Binomial distribution: for n independent Bernoulli trials $X_1, X_2, \ldots, X_n$, each with probability p of 'success', their sum $R = X_1 + X_2 + \ldots + X_n$ has a Binomial[p,n] distribution with mean np and variance $np(1 - p)$, where $\Pr(R = r) = \binom{n}{r} p^r (1 - p)^{n-r}$.

Brier score: a measure for the accuracy of probabilistic predictions, based on the mean squared predictive error. If $p_1, \ldots, p_n$ are the probabilities given to a set of n binary random variables $X_1, \ldots, X_n$ taking on values 0 and 1, then the usual Brier score is $\frac{1}{n}\sum_{i=1}^{n} (x_i - p_i)^2$. For a single outcome, this simplifies to $B = (1 - p)^2$, where p is the probability assigned to the event that occurs; the linear transformation $25-100B$ gives the scoring system shown in Table 2.3. For an outcome X_i with $K > 2$ categories, one of which is observed to be 1 and the rest 0, then given probabilities $p_{i1}, \ldots, p_{iK}$, the usual form of the Brier score for each observed event $x_{i1}, \ldots, x_{iK}$ is $\sum_{k=1}^{K} (x_{ik} - p_{ik})^2$.

Central limit theorem: the tendency for the sample mean to have a normal distribution, regardless of the shape of the underlying sampling distribution of the random variable. If n independent observations each have mean μ and variance σ^2, then under broad assumptions their sample mean is an estimator of μ, and has an approximately normal distribution with mean μ, variance σ^2/n and standard deviation $\sigma/\sqrt{n}$ (also known as standard error of the estimator).

Chance: (a) a general term for unavoidable unpredictability, (b) a generally agreed probability that can be assessed using physical characteristics, so that it can be loosely treated as an 'objective' property of the world.

Chaotic system: a completely deterministic system with non-linear relationships between events, in which very small changes in initial conditions can have major subsequent effects. This means that chaotic systems are so complex that they can be indistinguishable from stochastic systems in which there is genuine randomness.

Conditional independence: Random variables A and B are conditionally independent given a third variable C if $Pr(A,B|C) = Pr(A|C)Pr(B|C)$.

Conditional probability: The conditional probability of B given A is denoted $Pr(B|A)$, and defined as $Pr(B|A) = Pr(A \text{ and } B)/Pr(A)$.

Confidence: an informal term for a strength of belief; sometimes used in place of a numerical probability, and sometimes used to denote the quality of evidence behind an assessed probability.

Confidence interval: based on an observed set of data x, a 95% confidence interval for μ is an interval $(L(x), U(x))$ with the property that, before observing the data, there is a 95% probability that the random interval $(L(X), U(X))$ contains μ. The *central limit theorem*, combined with the knowledge that 95% of a normal distribution lies between the mean ± 2 standard deviations, means that a common approximation for a 95% confidence interval is the estimate ± 2 standard errors.

Cromwell's Rule: to always assign a small probability to all eventualities that are not logically impossible, after Oliver Cromwell's appeal to the Church of Scotland 'I beg you, in the bowels of Christ, think it possible you may be mistaken.' This can be considered a mathematical expression for humility.

Direct uncertainty: an expression of numerical or verbal uncertainty about an object. To be contrasted with *indirect uncertainty*.

Effectively random: a system that passes statistical tests for pure randomness, even if it is in fact deterministic.

Ensemble: a collection of computer runs with varying initial conditions or structure, producing a distribution of possible outcomes that are treated as equally plausible.

Epistemic uncertainty: lack of knowledge about facts, numbers or scientific hypotheses.

Excess fraction: see *attributable fraction*.

Exchangeable random variables: a set of random variables $X_1, \ldots, X_m$ is exchangeable if the joint probability $\Pr(X_1, \ldots, X_m)$ does not depend on the order of the variables. De Finetti's theorem says that exchangeable Bernoulli trials $X_1, \ldots, X_m$ have a joint distribution $\Pr(X_1, \ldots, X_m) = \prod_{i=1}^{m} \Pr(X_i|\theta)\Pr(\theta)d\theta$; i.e. the variables can be considered as conditionally independent given a parameter θ with a prior distribution $\Pr(\theta)$.

Expectation, mean: the mean average of a random variable. It is defined as $\Sigma x \Pr(X=x)$ for a discrete variable X and $\int x f(x)dx$ for a continuous variable.

Exponential distribution: A random variable X has an exponential distribution with mean m if its probability density is $f(x|m) = \frac{1}{m}e^{-x/m}$; $x > 0$, and $\Pr(X > x) = e^{-x/m}$; $x > 0$.

Exponential growth: when an initial quantity is multiplied by a fixed amount k in each unit of time. If we define $r = \log_e k$ (natural logarithms to base e), then this implies $k = e^r$, and after n units of time our initial quantity has increased by a factor e^{rn}.

F-N curves: a plot of the assessed cumulative frequency of events against their impact in terms of lives lost, both on a logarithmic (multiplicative) scale, so that a point (f, n) means that events with greater than n fatalities are assessed to occur at a frequency of f per year.

Generalized Pareto distribution (GPD): a variable X with a GPD with location μ, scale σ and shape a has a probability density $f(x|\mu, \sigma, a) = \frac{1}{\sigma}\left[1 + \frac{(x - \mu)}{a\sigma}\right]^{-(a+1)}$; $x \geq \mu$. The distribution function is $P(X \leq x|\mu, \sigma, a) = 1 - \left[1 + \frac{(x - \mu)}{a\sigma}\right]^{-a}$; $x \geq \mu$. When $\mu = a\sigma$, then the GPD is a Pareto distribution with shape a and minimum $x_m = a\sigma$.

Geometric distribution: Consider a set of independent Bernoulli trials, each with probability p. Then the attempt at which the first event occurs is a random variable X with a geometric distribution, with $\Pr(X = x) = (1-p)^{x-1}p$; $x = 1, 2, 3, \ldots$, and has expectation (mean average) $1/p$. For small p, the distribution is approximated by an exponential distribution with mean $1/p$, so that $\Pr(X > x \mid p) = e^{-xp}$.

Hazard: (a) a situation which presents the possibility of harm, but maybe in very extreme circumstances, (b) the 'instantaneous risk' in a defined period, such as the probability of dying in the next year, given survival up to now.

Hazard ratio: when analysing survival times, the hazard ratio associated with an exposure is the relative risk of suffering an event in a fixed period of time. A *Cox regression* is a form of multiple regression when the response variable is a survival time and the coefficients correspond to log(hazard ratios).

Independent events: A and B are independent if the occurrence of A does not influence the probability of B, so that $Pr(B|A) = Pr(B)$, or equivalently $Pr(B,A) = Pr(B)Pr(A)$.

Indirect uncertainty: a qualitative statement about the strengths and weaknesses of the evidence underlying a claim about a fact, including gaps in potentially influential information.

Likelihood: technically, a measure of the evidential support provided by data for particular parameter values. When a probability distribution for a random variable depends on a parameter, say θ, then after observing data x the likelihood for θ is proportional to $Pr(x|\theta)$.

Likelihood ratio: a measure of the relative support that some data provides for two competing hypotheses. For hypotheses H_0 and H_1, the likelihood ratio provided by data x is given by $Pr(x|H_0)/Pr(x|H_1)$.

Markov Chain Monte Carlo (MCMC): a sequential simulation procedure for Bayesian analysis of a complex statistical model. Unknown parameters and other variables are sampled from distributions that are conditional solely on the immediately preceding values, thus forming a Markov chain. An appropriate choice of the sampling method leads to convergence of the sampled values to their correct posterior distribution.

Mean (of a population): see *expectation*.

Mean (of a sample): Suppose we have a set of n data points, which we label as $x_1, x_2, \ldots, x_n$. Then their *sample mean* is given by $m = (x_1 + x_2 + \cdots x_n)/n$. For example, if 3, 2, 1, 0, 1 are the numbers of children reported by 5 people in a sample, then the sample mean is $(3+2+1+0+1)/5 = 7/5 = 1.4$.

Median (of a sample): The *sample median* is the value midway along an ordered set of data points: if n is odd, then the sample median is the middle value; if n is even, then the average of the two 'middle' points is taken as the median.

Mode (of a population distribution): the response with the maximum probability of occurring.

Mode (of a sample): the most common value in a set of data.

Model: see *statistical model*.

Monte Carlo methods: rather than attempting a mathematical analysis of a statistical model, a Monte Carlo analysis simulates a series of examples of the unknown random variables and calculates any quantities of interest. This produces an empirical distribution of any outputs, which can then be summarized and reported. Also known as *stochastic simulation*.

Normal or Gaussian distribution: X has a normal (Gaussian) distribution with mean μ and variance σ^2 if it has a probability density function $f(x|\mu, \sigma^2) = \frac{1}{\sqrt{2\pi\sigma^2}} e^{-\frac{(x-\mu)^2}{2\sigma^2}}$, for $-\infty \leq x \leq \infty$. Then $E(X) = \mu$, $V(X) = \sigma^2$, $SD(X) = \sigma$.

Null hypothesis: a default scientific theory, generally representing the absence of an effect or finding of interest, which is tested using a P-value. Generally denoted H_0.

Odds: If any event has probability p, then the odds are $p/(1-p)$.

Ontological uncertainty: when we are uncertain about the whole conceptualization of a problem; what the possibilities are, what the important influential factors are, what the terms mean. Essentially, what are we actually dealing with?

P-value: a measure of discrepancy between data and a null hypothesis. For a null hypothesis H_0, let T be a test statistic for which large values indicate inconsistency with H_0. Suppose we observe a value t. Then a (one-sided) P-value is the probability of observing such an extreme value, were H_0 true, that is $Pr(T \geq t|H_0)$. If both small and large values of T indicate inconsistency with H_0, then the two-sided P-value is the probability of observing such an extreme value in either direction, and different definitions can be made of 'extreme'. Often the two-sided P-value is simply taken as double the one-sided P-value: the R software uses the total probability of events which have a lower probability of occurring than what was observed.

Pareto distribution: a probability distribution for a random variable X with density $Pr(x|x_m, a) = (a/x_m)/(x/x_m)^{a+1}; x > x_m$ where $Pr(X > x|x_m, a) = 1/(x/x_m)^a; x > x_m$.

Poisson distribution: a distribution for a random variable X with parameter μ and $Pr(X = x|\mu) = e^{-\mu}\frac{\mu^x}{x!}$ for $x = 0, 1, 2, \ldots$. Then $E(X) = \mu$ and $V(X) = \mu$.

Polygenic risk scores: using genetic data from an individual to assess the probability of various adverse health events occurring in the future. Crucial issues include whether they add much predictive value above other, more routine data, and what the impact might be when individuals are told their risk scores.

Population attributable fraction: If P is the prevalence of the causal risk factor with a relative risk RR, then the population attributable fraction is $PAF = P(RR - 1)/(P(RR - 1) + 1)$.

Posterior probability/distribution/odds: in Bayesian analysis, the probability distribution of unknown quantities A after taking into account observed data B through Bayes' theorem to give $Pr(A|B)$. The posterior odds for a proposition A would be $Pr(A|B)/Pr(\text{not } A|B)$.

Power law: see *Pareto distribution*.

Prior probability/distribution/odds: in Bayesian analysis, the initial probability given to the quantity of interest, before being updated into a posterior distribution by the data.

Probability: Let $Pr(A)$ be the probability for an event A. Then

1. *Bounds*: $0 \leq Pr(A) \leq 1$, with $Pr(A) = 0$ if A is impossible and $Pr(A) = 1$ if A is certain.
2. *Complement*: $Pr(A) = 1 - Pr(\text{not } A)$.
3. *Addition rule*: If A and B are mutually exclusive (i.e. one at most can occur), $Pr(A \text{ or } B) = Pr(A) + Pr(B)$.
4. *Multiplication rule*: for any events A and B, $Pr(A \text{ and } B) = Pr(A|B)$ $Pr(B)$, where $Pr(A|B)$ represents the probability for A *given* B has occurred. A and B are independent if and only if $Pr(A|B) = Pr(A)$, i.e. the occurrence of B does not affect the probability for A. In this case we have $Pr(A \text{ and } B) = Pr(A)Pr(B)$, the multiplication rule for independent events.

Probability distribution: a generic term for a mathematical expression of the probability of a random variable taking on particular values. A random variable X has a probability distribution function defined by $F(x) = Pr(X \leq x)$, for all $-\infty < x < \infty$, *i.e.* the probability that X is at most x. A discrete random variable can take only a limited set of values, and it takes on the value x with probability $Pr(X = x)$. A continuous random variable X has a probability density function $f(x)$ such that $P(X \leq x) = \int_{-\infty}^{x} f(t)\,dt$, and expectation given by $E(X) = \int_{-\infty}^{\infty} x f(x)\,dx$. The probability of X lying in the interval (A,B) can be calculated using $\int_{A}^{B} f(x)\,dx$.

Proportional hazard model: a statistical model for survival analysis in which there is a baseline function $h_0(t\,;x_0)$ specifying the hazard at time t for a case with covariates x_0. For a case with covariates x_1, the hazard function becomes $h_1(t\,;x_1) = h_0(t\,;x_0) \times \lambda_{x1-x0}$

Prosecutor's fallacy: when $\Pr(A|B)$ is mistakenly interpreted as $\Pr(B|A)$. Also known as the 'transposed conditional'.

Pseudo-random number generator: an algorithm for producing numbers that are statistically indistinguishable from a random sequence, but are generated using a wholly deterministic process.

Quantitative risk analysis: a process in which potential outcomes are listed, and their probabilities are assessed using statistical models. Values for the outcomes may also be assigned.

Random variable: a quantity with a probability distribution (more formally it is a mapping from the space of all possible outcomes to a number). Before they are observed, random variables are generally given capital letters.

Relative risk: if the absolute risk among people who are not exposed to something of interest is p, and the absolute risk among people who are exposed is q, then the *relative risk* is q/p.

Risk: a term with a wide variety of uses, from potential threats to the probability of something bad happening.

Standard deviation: the square root of the variance, $SD(X) = \sqrt{V(X)}$. For well-behaved, reasonably symmetric data distributions without long tails, we would expect most of the observations to lie within two standard deviations from the mean.

Statistical model: a mathematical representation, containing unknown parameters, of the probability distribution of a set of random variables, often involving complex relationships.

Stochastic: when there is unavoidable randomness in a non-deterministic system, meaning that, in principle, two completely identical situations could end up with different outcomes.

Uncertainty interval: a generic name for an interval around an estimate of a particular unknown quantity, often labelled as, say, '95%'. These could be confidence intervals, in which case 95% of such intervals will contain the true value if all the modelling assumptions hold, or credible intervals calculated from a Bayesian analysis with 95% probability of containing the true value if the modelling assumptions hold, or an informal judgement assigned 95% subjective probability of containing the true value.

Uniform distribution: a continuous random variable X has a uniform distribution on the interval $[a,b]$ if its density is $f(x|a, b) = 1/(b - a)$; $a \leq x \leq b$. A discrete random variable X has a uniform distribution on say the integers $1, \ldots, n$ if $\Pr(X = x|n) = 1/n; x = 1, \ldots, n$.

Variance: the variance of a random variable X is $V(X) = E((X - E(X))^2)$.

Variance of a sample: this is generally defined as $s^2 = \frac{1}{(n-1)} \sum_{i=1}^{n} (x_i - \bar{x})^2$, although setting the denominator to be n is not incorrect and can be justified.

Index